THE FOURTH REVOLUTION

THE 4TH REVOLUTION

HOW THE INFOSPHERE IS RESHAPING HUMAN REALITY

LUCIANO FLORIDI

OXFORD
UNIVERSITY PRESS

OXFORD

UNIVERSITY PRESS

Great Clarendon Street, Oxford, OX2 6DP,
United Kingdom

Oxford University Press is a department of the University of Oxford.
It furthers the University's objective of excellence in research, scholarship,
and education by publishing worldwide. Oxford is a registered trade mark of
Oxford University Press in the UK and in certain other countries

First Edition published in 2014

Impression: 1

Published in the United States of America by Oxford University Press
198 Madison Avenue, New York, NY 10016, United States of America

British Library Cataloguing in Publication Data
Data available

Library of Congress Control Number: 2013957566

ISBN 978–0–19–960672–6

Printed in Italy by
L.E.G.O. S.p.A.

CONTENTS

PREFACE

This book is about how our digital ICTs (information and communication technologies) are affecting our sense of self, how we relate to each other, and how we shape and interact with our world. Nanotechnology, the Internet of Things, Web 2.0, Semantic Web, cloud computing, motion-capturing games, smartphone apps, tablets and touch screens, GPS, Augmented Reality, artificial companions, unmanned drones, driverless cars, wearable computing devices, 3D printers, identity theft, online courses, social media, cyberwar...the technophile and the technophobe ask the same question: what's next? The philosopher wonders what lies behind. Is there a unifying perspective from which all these phenomena may be interpreted as aspects of a single, macroscopic trend? Part of the difficulty, in answering this question, is that we are still used to looking at ICTs as tools for interacting with the world and with each other. In fact, they have become environmental, anthropological, social, and interpretative forces. They are creating and shaping our intellectual and physical realities, changing our self-understanding, modifying how we relate to each other and ourselves, and upgrading how we interpret the world, and all this pervasively, profoundly, and relentlessly.

So this is a philosophical book, yet it is not a book just for philosophers. It seeks to identify and explain some of the deep technological forces that are affecting our lives, our beliefs, and anything that surrounds us, but it is not a technical or scholarly treatise. As the reader will notice by quickly browsing the Contents, I believe we are seeing the beginning a profound cultural revolution, largely driven by ICTs. I know that every generation thinks it is special just because it is alive and hence uniquely placed, reflectively, between the dead and the unborn. So I agree that it is important to keep things in perspective. However, sometimes it is 16 December 1773 and you are in Boston, or it is 14 July 1789 and you are in Paris. What I stress in this book is that sometimes it is a new millennium, and you are in the infosphere.

The information revolution that I discuss is a great opportunity for our future. So this is also a moderately optimistic book. I say 'moderately' because the question is whether we shall be able to make the most of our ICTs, while avoiding their worst consequences. How can we ensure that we shall reap their benefits? What could we do in order to identify, coordinate, and foster the best technological transformations? What are the risks implicit in transforming the world into a progressively ICT-friendly environment? Are our technologies going to enable and empower us, or will they constrain our physical and conceptual spaces, and quietly force us to adjust to them because that is the best, or sometimes the only, way to make things work? Can ICTs help us to solve our most pressing social and environmental problems, or are they going to exacerbate them? These are only some of the challenging questions that the information revolution is posing. My hope is that this book may contribute to the larger ongoing effort to clarify and address them; and that a more fruitful and effective approach to the problems and opportunities of ICTs may be possible, if we gain a deeper and more insightful understanding of their impact on our current and future lives.

The great opportunity offered by ICTs comes with a huge intellectual responsibility to understand them and take advantage of them in the right way. That is also why this is not a book for specialists but for everyone who cares about the development of our technologies and how they are affecting us and humanity's foreseeable future. The book does not presuppose any previous knowledge of the topics, even if it is not an elementary text for beginners. Complex phenomena can be simplified conceptually, but there is a threshold beyond which the simplification becomes an unreliable and therefore useless distortion. I have tried to walk as closely as possible to that threshold without crossing it. I hope the reader will judge my efforts kindly.

As a book for non-specialists, it may double as an introduction. For it is part of a wider project, on the foundations of the philosophy of information, which seeks to update our philosophy, and make it relevant to our time and beyond academic walls.[1] Given the unprecedented novelties that the dawn of the information era is producing, it is not surprising that many of our fundamental philosophical views, so entrenched in history and above all in the industrial age, may need to be upgraded and complemented, if not entirely replaced. Perhaps not yet in academia, think tanks, research centres, or R & D offices, but clearly in the streets and online, there is an atmosphere of confused expectancy mixed with concern; an awareness of exciting, bottom-up changes occurring in our views about the world, ourselves, and our interactions with the world and with each other. This atmosphere and this awareness are not the result of research programmes, or of the impact of successful grant applications. Much more realistically and powerfully, but also more confusedly and tentatively, the alterations in our views of the world are the result of our daily adjustments, intellectually and behaviourally, to a reality that is fluidly changing in front of our eyes and under our feet, exponentially and unremittingly. We are finding a new balance as we rush into the future, by shaping and adapting to new conditions that have not yet become sedimented into maturity. Novelties no longer result in initial disruption fading into finally stable patterns of 'more of approximately the same'. Think, for example, of the car or the book industry, and the stability they ended up providing, after an initial period of disruptions and rapid adjustments. It seems clear that a new philosophy of history, which tries to make sense of our age as the end of history and the beginning of hyperhistory (more on this concept in Chapter 1), invites the development of a new philosophy of nature, a new philosophical anthropology, a synthetic environmentalism as a bridge between us and the world, and a new philosophy of politics among us. 'Cyberculture', 'posthumanism', 'singularity', and other similar fashionable ideas can all be understood as attempts to make sense of our new hyperhistorical predicament. I find them indicative and sometimes suggestive, even if unconvincing. 'O buraco é mais embaixo', as they say in Brazil: the hole is way deeper, the problem much more profound. We need to do some serious philosophical digging. This is why the invitation to rethink the present and the future in an increasingly technologized world amounts to a request for a new philosophy of information that can apply to every aspect of our hyperhistorical condition. We need to look carefully at the roots of our culture and nurture them, precisely because we are rightly concerned with its leaves and flowers.

We know that the information society has its distant roots in the invention of writing, printing, and the mass media. However, it became a reality only recently, once the *recording* and *transmitting* facilities of ICTs evolved into *processing* capabilities. The profound and widespread transformations brought about by ICTs have caused a huge conceptual deficit. We clearly need philosophy to be on board and engaged, for the tasks ahead are serious. We need philosophy to grasp better the nature of information itself. We need philosophy to anticipate and steer the ethical impact of ICTs on us and on our environments. We need philosophy to improve the economic, social, and political dynamics of information. And we need philosophy to develop the right intellectual framework that can help us semanticize (give meaning to and make sense of) our new predicament. In short, we need a philosophy of information as a philosophy *of* our time *for* our time.

I have no illusions about the gigantic task ahead of us. In this book, I only sketch a few ideas for a philosophy of history, in terms of a philosophy of hyperhistory; for a philosophy of nature, in terms of a philosophy of the infosphere; for a philosophical anthropology, in terms of a fourth revolution in our self-understanding, after the Copernican, the Darwinian, and Freudian ones; and for a philosophy of politics, in terms of the design of multi-agent systems that may be up to the task of dealing with global issues. All this should lead to an expansion of ethical concerns and care for all environments, including those that are artificial, digital, or synthetic. Such a new 'e-nvironmental' ethics should be based on an information ethics for the whole infosphere and all its components and inhabitants. In the following chapters, I only touch upon such ideas and outline the need for an ethical infrastructure that may be coherent with them. Much more work lies ahead. I very much hope that many others will be willing to join forces.

Finally, the reader will see that this book contains plenty of terminology that is only tentative, with neologisms, acronyms, and technical expressions. Similar attempts to reshape our language can be vexing, but they are not always avoidable. The struggle to find a balance between readability and accuracy is obvious and I decided not to hide it. To rephrase a colourful analogy by Friedrich Waismann (1896–1959), a philosopher member of the Vienna Circle, just as a good swimmer is able to swim upstream, so a good philosopher may be supposed to be able to master the difficult art of thinking 'up-speech', against the current of linguistic habits.[2] I fully agree, but I am also aware that my efforts to capture the profound intellectual novelties that we are facing remain inadequate. The challenge of withstanding the flow of old ideas is serious, because there can hardly be better policies without a better understanding. We may need to reconsider and redesign our conceptual vocabulary and our ways of giving meaning to, and making sense of, the world (our semanticizing processes and practices) in order to gain a better grasp of our age, and hence a better chance to shape it in the best way and deal successfully with its open problems. At the same time, this is no licence to give up clarity and reason, relevant evidence and cogent arguments, plausible explanations, and honest admissions of uncertainty or ignorance. Swimming against the current is not equivalent to splashing around in panic. On the contrary, discipline becomes even more essential. We need to improve our intellectual condition, not give it up. So perhaps I may adapt another aquatic metaphor,[3] introduced this time by Otto Neurath (1882–1945), also a philosopher member of the Vienna Circle: we do not even have a raft, but drowning in obscurities is not an option.[4] Lazy thinking will only exacerbate our problems. We need to make a rational effort and build a raft while still swimming. I hope the following chapters provide some timber.

ACKNOWLEDGEMENTS

So many people helped me when writing this book, in so many ways, and on so many occasions, that I am sure that if I were to try to name all of them I would still forget to mention someone important, no matter how long the list could be. So I shall limit myself to thank only those who have been more influential in the last stage of the research and of the writing marathon.

I am tremendously grateful to Latha Menon, senior commissioning editor at OUP, for having encouraged me to commit myself to this ambitious project, for her input at several stages of the work, and for her support throughout the years, even when I kept asking for more deadline extensions. She read the penultimate draft and made it remarkably more reader-friendly.

Many conversations with Anthony Beavers, Terry Bynum, Massimo Durante, Charles Ess, Amos Golan, Mireille Hildebrandt, Hosuk Lee-Makiyama, Marco Pancini, Ugo Pagallo, Mariarosara Taddeo, Matteo Turilli, Menno van Doorn, and Marty J. Wolf on different parts of this book led to major improvements. We did not waste our wine, but I still owe them several drinks. In particular, Massimo Durante, Federico Gobbo, Carson Grubaugh, Ugo Pagallo, and Marty J. Wolf read what I thought was the last draft and transformed it into a penultimate one, thanks to their very insightful feedback.

I owe to my wife, Anna Christina (Kia) De Ozorio Nobre, not only a life full of love, but also the initial idea of dedicating more attention to the 'fourth revolution', and a boundless faith in her husband's abilities to live up to her high expectations and standards. She heard me complaining so often about how difficult it was to complete this book that I am almost ashamed I actually managed it. Few things motivate as much as the complete certainty, in someone you love and esteem, that you will succeed. Kia made many essential and insightful suggestions about the last draft, which I read to her in the course of some wonderful evenings in front of our fireplace.

In 2012, I had the pleasure and privilege to chair a research group, the Onlife Initiative, organized by the European Commission, on the impact of ICTs on the digital transformations occurring in the European society. Nicole Dewandre, adviser to the Director-General, Directorate General for Communications Networks, Content and Technology of the European Commission, initiated and strongly supported the whole project, and I am deeply indebted to her and to Robert Madelin for such a wonderful challenge to do some philosophy in the real world. The output of the group's activities was *The Onlife Manifesto*.[1] It was a great honour to have the group and the manifesto named after some of the ideas I present in this book. Being part of the group was an amazing intellectual experience. Through it, I came to understand better many aspects of the information revolution that I would have probably missed without the input and the conversation of so many exceptional colleagues. So, many thanks to my fellow 'onlifers': Franco Accordino, Stefana Broadbent, Nicole Dewandre, Charles Ess, Jean-Gabriel Ganascia, Mireille Hildebrandt, Yiannis Laouris, Claire Lobet, Sarah Oates, Ugo Pagallo, Judith Simon, May Thorseth, and Peter Paul Verbeek.

The final book is the result of the very fruitful interactions I enjoyed with the OUP's editorial team, and in particular with Emma Ma. The anonymous reviewers appointed by OUP kept me on the right path. Penny Driscoll, my personal assistant, skilfully proofread the manuscript, making it much more readable. She also provided some very helpful philosophical feedback on the final version of the book. I must confirm here what I have already written before: without her exceptional support and impeccable managerial skills I could not have completed this project.

Finally, I would like to thank the University of Hertfordshire, Brendan Larvor, and Jeremy Ridgman for having provided me with all the support necessary to pursue my research at different stages during the past few years; the British Arts and Humanities Research Council and Google, for three academic grants, during the academic years 2010/11 and 2011/12, that supported some of the research for this book; Amos Golan, who kindly invited me to join, as adjunct professor, the Infometrics Institute at the Department of Economics of the American University (AU) in Washington; and my recent new academic home, the Oxford Internet Institute. The last writing effort was made possible thanks to a quiet, focused, and systematic period of time that I had the privilege to spend at AU in 2013.

LIST OF FIGURES

1

TIME

Hyperhistory

The three ages of human development

More people are alive today than ever before in human history. And more of us live longer today than ever before. Life expectancy is increasing (Figure 1; see also Figure 19) and poverty is decreasing (Figure 2), even if the degree of global inequality is still scandalous. As a result, disability is becoming the biggest health-related issue for humanity.

To a large measure, the lines representing the trends on Figure 1 and Figure 2 have been drawn by our technologies, at least insofar as we develop and use them intelligently, peacefully, and sustainably.

Sometimes we forget how much we owe to flints and wheels, to sparks and ploughs, to engines and computers. We are reminded of our deep technological debt when we divide human life into *prehistory* and *history*. Such a significant threshold is there to acknowledge that it was the invention and development of ICTs (information and communication technologies) that made all the difference between who we were, who we are, and, as I shall argue in this book, who we could be and become. It is only when systems to record events and hence accumulate and transmit information for future consumption became available that lessons learnt by past generations began to evolve exponentially, in a soft or Lamarckian[1] way, and so humanity entered into history.

1

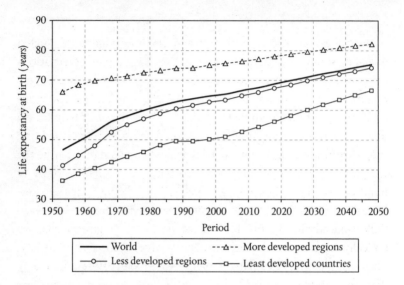

Fig. 1. Life Expectancy at Birth for the World and Major Development Group, 1950–2050.

Source: Population Division of the Department of Economic and Social Affairs of the United Nations Secretariat (2005). *World Population Prospects: The 2004 Revision Highlights.* New York: United Nations.

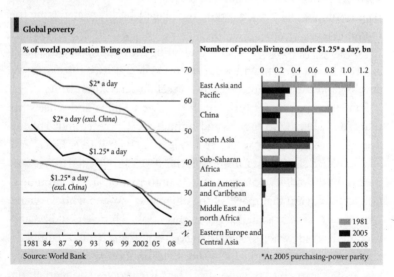

Fig. 2. Poverty in the World Defined as the Number and Share of People Living Below $1.25 a Day (at 2005 prices) in 2005–2008.

Source: World Bank. © *The Economist* Newspaper Limited, London (29 February 2012).

History is therefore synonymous with the information age. Such a line of reasoning may suggest that humanity has been living in various kinds of information societies at least since the Bronze Age, the era that marks the invention of writing in Mesopotamia and other regions of the world (4th millennium BC). Indeed, in the 3rd millennium BC, Ur, the city state in Sumer (Iraq), represented the most developed and centralized bureaucratic state in the world. So much so that, before the Gulf War (1991) and the Iraq War (2003–11), we still had a library of hundreds of thousands of clay tablets. They contain neither love letters nor holiday stories, but mainly inventories, business transactions, and administration documents. And yet, Ur is not what we typically have in mind when we speak of an information society. There may be many explanations, but one seems more convincing than any other: only very recently has human progress and welfare begun to be not just *related to*, but *mostly dependent on*, the successful and efficient management of the life cycle of information. I shall say more about such a cycle in the rest of this chapter, but, first, let us see why such a dependency has meant that we recently entered into *hyperhistory* (Figure 3).

Prehistory and history work like adverbs: they tell us *how* people live, not *when* or *where* they live. From this perspective, human societies currently stretch across three ages, as ways of living. According

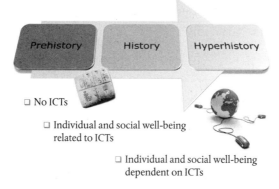

Fig. 3. From Prehistory to Hyperhistory.

to reports about an unspecified number of uncontacted tribes in the Amazonian region,[2] at the beginning of the second millennium there were still some societies that may be living prehistorically, without recorded documents. If, or rather when, one day such tribes disappear, the end of the first chapter of our evolutionary book will have been written.

The greatest majority of people today still live historically, in societies that rely on ICTs to record, transmit, and use data of all kinds. In such historical societies, ICTs have not yet overtaken other technologies, especially energy-related ones, in terms of their vital importance. Then, there are some people around the world who are already living hyperhistorically, in societies and environments where ICTs and their data-processing capabilities are not just important but essential conditions for the maintenance and any further development of societal welfare, personal well-being, and overall flourishing. For example, all members of the G7 group—namely Canada, France, Germany, Italy, Japan, the United Kingdom, and the United States of America—qualify as hyperhistorical societies because, in each country, at least 70 per cent of the Gross Domestic Product (GDP, the value of goods and services produced in a country) depends on intangible goods, which are information-related, rather than on material goods, which are the physical output of agricultural or manufacturing processes. Their economies heavily rely on information-based assets (knowledge-based economy), information-intensive services (especially business and property services, communications, finance, insurance, and entertainment), and information-oriented public sectors (especially education, public administration, and health care).

The nature of conflicts provides a sad test for the reliability of this tripartite interpretation of human evolution. Only a society that lives hyperhistorically can be threatened informationally, by a cyber attack. Only those who live by the digit may die by the digit, as we shall see in Chapter 8.

Let us return to Ur. The reason why we do not consider Ur an information society is because it was historical but not yet hyperhistorical. It depended more on agricultural technologies, for example, than on clay tablets. Sumerian ICTs provided the recording and transmitting infrastructure that made the escalation of other technologies possible, with the direct consequence of furthering our dependence on more and more layers of technologies. However, the recording and transmitting facilities of ICTs evolved into processing capabilities only millennia later, in the few centuries between Johann Gutenberg (c.1400–68) and Alan Turing (1912–54). It is only the present generation that is experiencing the radical transformations, brought about by ICTs, which are drawing the new threshold between history and hyperhistory.

The length of time that the evolution of ICTs has taken to bring about hyperhistorical information societies should not be surprising. The life cycle of information (see Figure 4) typically includes the

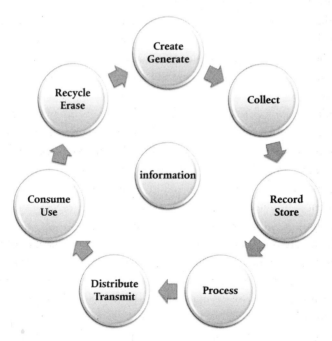

Fig. 4. A Typical Life Cycle for Information.

following phases: *occurrence* (discovering, designing, authoring, etc.), *recording, transmission* (networking, distributing, accessing, retrieving, etc.), *processing* (collecting, validating, merging, modifying, organizing, indexing, classifying, filtering, updating, sorting, storing, etc.), and *usage* (monitoring, modelling, analysing, explaining, planning, forecasting, decision-making, instructing, educating, learning, playing, etc.). Now, imagine Figure 4 to be like a clock and a historian writing in the future, say in a million years. She may consider it normal, and perhaps even elegantly symmetrical, that it took roughly 6,000 years for the agricultural revolution to produce its full effect, from its beginning in the Neolithic (10th millennium BC), until the Bronze Age, and then another 6,000 years for the information revolution to bear its main fruit, from the Bronze Age until the end of the 2nd millennium AD. She may find it useful to visualize human evolution as a three-stage rocket: in prehistory, there are no ICTs; in history, there are ICTs, they record and transmit information, but human societies depend mainly on other kinds of technologies concerning primary resources and energy; and in hyperhistory, there are ICTs, they record, transmit, and, above all, process information, increasingly autonomously, and human societies become vitally dependent on them and on information as a fundamental resource in order to flourish. Around the beginning of the 3rd millennium, our future historian may conclude, innovation, welfare, and added value moved from being ICT-related to being ICT-dependent. She might suppose that such a shift required unprecedented levels of processing power and huge quantities of data. And she might suspect that memory and connectivity must have represented bottlenecks of some kind. She would be right on both accounts, as we shall see in the rest of this chapter.

Instructions

Consider the two diagrams in Figure 5 and Figure 6. Figure 5 is famous, almost iconic. It is known as Moore's Law and suggests that, over the

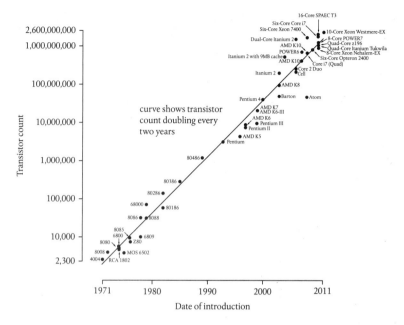

Fig. 5. Moore's Law.
Source: Wikipedia.

period of development of digital computers, the number of transistors on integrated circuits doubles approximately every two years.

Figure 6 is less famous but equally astonishing. It tells you a similar story, but in terms of decreasing cost of computational power. In 2010, an iPad2 had enough computing power to process 1,600 millions of instructions per second (MIPS). By making the price of such a processing power equal to $100, the graph shows what it would have cost to buy the computing power of an iPad2 in the past decades. Note that the vertical scale is logarithmic, so it descends by powers of ten as the price of computing power decreases dramatically. All this means that, in the fifties, the 1,600 MIPS you hold in your hands—or rather held, in 2010, because three years later the iPad4 already run at 17,056 MIPS—would have cost you $100 trillion. This is a number that only bankers and generals understand. So, for a quick comparison, consider

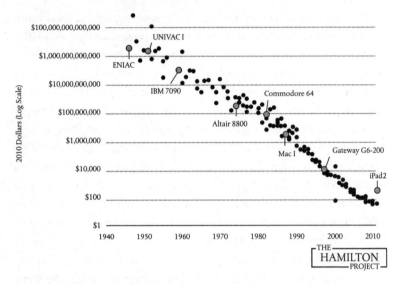

Fig. 6. The Cost of Computing Power Equal to an iPad2.
Source: The Hamilton Project at the Brookings Institution.

Qatar's GDP. In 2010, it was ranked 57th out of 190 countries in the world and its GDP would have been insufficient to buy the equivalent of an iPad2 in the fifties, for it was a mere $98 trillion.

Whether you find Figure 5 or Figure 6 more compelling, the conclusion is the same: increasingly more power is available at decreasing costs, to ever more people, in quantities and at a pace that are mind-boggling. The limits of computing power seem to be mainly physical. They concern how well our ICTs can dissipate heat and recover from unavoidable hardware faults while becoming increasingly small. This is the rocket that has made humanity travel from history to hyperhistory, to use a previous analogy. It also explains why ICTs are still disruptive technologies that have not sedimented: new generations keep teaching the old ones how to use them, although they still learn from previous generations how to drive or use a microwave.

At this point, an obvious question is where all this computational power goes. It is not that we are regularly putting people on the Moon with our smartphones or tablets. The answer is: interactions, both machine-to-machine and human–computer ones, also known as HCI.

In machine-to-machine interactions, an ICT system, such as a meter or sensor, monitors and records an event, such as the state of the road surface, and communicates the resulting data through a network to an application, which processes the data and acts on them, for example by automatically adapting the braking process of a car, if necessary. You might have heard that there is more computational power in an average new car today than was available to NASA to send astronauts to the Moon (Apollo mission, 1969). It is true. There are more than 50 ICT systems in an ordinary car, controlling anything from satellite navigation to hi-fi display, from ABS (anti-locking brakes) to electric locks, from entertainment systems to all the sensors embedded in the engine. It is a growing market in the automobile industry, as Figure 7 illustrates. According to Intel, the connected car is already the third fastest growing technological device after phones and tablets. It is only a matter of (short) time before all new cars will be connected to the Internet and, for example, find a convenient car park space, sense other vehicles, or spot cheaper petrol prices along the journey. And of course electric vehicles will require more and more 'computation': by 2015, they will contain about twice as many semiconductors than conventional cars. Mechanics are becoming computer engineers.

In human–computer interactions (HCI), ICTs are used to create, facilitate, and improve communications between human users and computational systems. When talking about ICTs, it is easy to forget that computers do not compute and telephones do not phone, to put it slightly paradoxically. What computers, smartphones, tablets, and all the other incarnations of ICTs do is to handle data. We rely on their capacities to manage huge quantities of MIPS much less to add numbers or call our friends than to update our Facebook status, order and read the latest e-books online, bill someone, buy an airline ticket, scan an electronic boarding pass, watch a movie, monitor the inside of a shop, drive to a place, or, indeed, almost anything else. This is why HCI is so important. Indeed, since the mid-1990s, HCI does not even

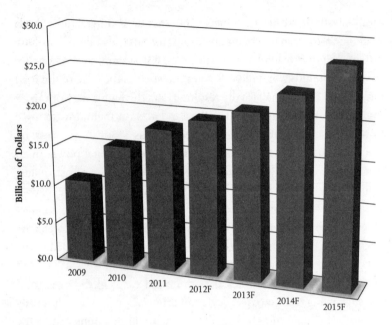

Fig. 7. Value in US Dollars of the Average Semiconductor Content in Automobiles. F=forecast.

Data source: IC Insights, 2012.

have to involve screens or keyboards. It may be a matter of a neuro-prosthetic device implanted in the brain. Of course, in all human–computer interactions, the better the process, the computationally greedier the ICT in question is likely to be. It takes a lot of MIPS to make things easy. This is the reason why new operating systems can hardly run on old computers.

We know that what our eyes can see in the world—the visible spectrum of the rainbow—is but a very small portion of the electromagnetic spectrum, which includes gamma rays, X-rays, ultraviolet, infrared, microwaves, and radio waves. Likewise, the data processing 'spectrum' that we can perceive is almost negligible compared to what is really going on in machine-to-machine and human–computer interactions. An immense number of ICT applications run an incalculable number of instructions every millisecond of our lives to keep the hyperhistorical information society humming. ICTs consume most of their MIPS to talk to each other, collaborate, and coordinate efforts, and put us as comfortably as possible in or on the loop, or even out of it, when necessary. According to a recent White Paper published by CISCO IBSG,[3] a multinational corporation that admittedly designs, manufactures, and sells networking equipment, there will be 25 billion devices connected to the Internet by 2015 and 50 billion by 2020 (see Figure 8).

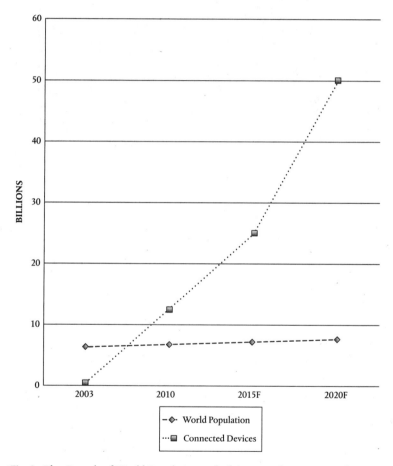

Fig. 8. The Growth of World Population and of Connected Devices. F = forecast. *Data source*: Evans (2011).

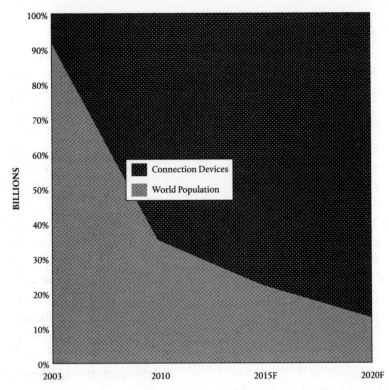

Fig. 9. The Total Space of Connectivity in Relation to the Growth of World Population and of Connected Devices. F = forecast.

Data source: Evans (2011).

The number of connected devices per person will grow from 0.08 in 2003, to 1.84 in 2010, to 3.47 in 2015, to 6.58 in 2020. To our future historian, global communication on Earth will appear to be largely a non-human phenomenon, as Figure 9 illustrates.

Almost all MIPS are invisible to us, like the oxygen we breathe, but they are becoming almost as vital, and they are growing exponentially. Computational devices of all sorts generate a staggering amount of data, much more data than humanity has ever seen in its entire history (Figure 10). This is the other resource that has made hyperhistory possible: zettabytes.

Data

A few years ago, researchers at Berkeley's School of Information[4] estimated that humanity had accumulated approximately 12 exabytes[5] of data in the course of its entire history until the commodification of computers, but that it had already reached 180 exabytes by 2006. According to a more recent study,[6] the total grew to over 1,600 exabytes between 2006 and 2011, thus passing the zettabyte (1,000 exabytes) barrier. This figure is now expected to grow fourfold approximately every three years, so that we shall have 8 zettabytes of data by 2015. Every day, enough new data are being generated to fill all US libraries eight times over. Of course, armies of ICT devices are constantly working to keep us afloat and navigate through such an ocean of data. These are all numbers that will keep growing quickly and steadily for the foreseeable future, especially because those very devices are among the greatest sources of further data, which in turn require, or simply make possible, more ICTs. It is a self-reinforcing cycle and it would be unnatural not to feel overwhelmed. It is, or at least should be, a mixed feeling of apprehension about the risks, excitement at the opportunities, and astonishment at the achievements, as we shall see in the following chapters.

Thanks to ICTs, we have entered *the age of the zettabyte*. Our generation is the first to experience a zettaflood, to introduce a neologism to describe this tsunami of bytes that is submerging our environments. In other contexts, this is also known as 'big data' (Figure 10).

Despite the importance of the phenomenon, it is unclear what exactly the term 'big data' means. The temptation, in similar cases, is to adopt the approach pioneered by Potter Stewart, United States Supreme Court Justice, when asked to describe pornography: difficult to define, but 'I know when I see it'. Other strategies have been much less successful. For example, in the United States, the National Science Foundation (NSF) and the National Institutes of Health (NIH) have identified big data as a programme focus. One of the main NSF–NIH

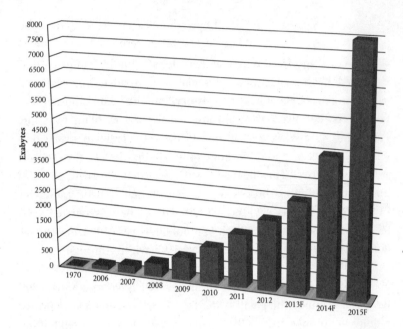

Fig. 10. The Growth of Big Data.

Source: based on IDC white paper, 'The Diverse and Exploding Digital Universe', March 2008, and IDC white paper, 'Worldwide Big Data Technology and Service 2012–2015 Forecast', March 2012.

interagency initiatives addresses the need for core techniques and technologies for advancing big data science and engineering. However, the two agencies specify that

> The phrase 'big data' in this solicitation refers to large, diverse, complex, longitudinal, and/or distributed data sets generated from instruments, sensors, Internet transactions, email, video, click streams, and/or all other digital sources available today and in the future.[7]

You do not need to be a logician to find this both obscure and vague. Wikipedia, for once, is also unhelpful. Not because the relevant entry is unreliable, but because it reports the common definition, which describes 'big data' as a collection of data sets so large and complex that it becomes difficult to process using available management tools or traditional data-processing applications. Apart from the circular problem of defining 'big' with 'large' (the NSF and NHI seem to be happy with it), understanding 'big data' in terms of 'small tools' suggests that data are too big or large only in relation to our current computational power. This is misleading. Of course, 'big', as many other terms, is a relational predicate: a pair of shoes may be too big for you, but fine for me. It is also trivial to acknowledge that we tend to evaluate things non-relationally, in this case as absolutely big, whenever the frame of reference is obvious enough to be left implicit. A horse is a big animal, no matter what whales may think. Yet these two simple points may give the impression that there is no real trouble with 'big data' being a loosely defined term referring to the fact that our current computers cannot handle so many gazillions of data efficiently. And this is where two confusions seem to creep in. First, that the *epistemological* (that is, knowledge-related) *problem* with big data is that there is too much of it (the *ethical problem* concerns how we use them, more on this presently). And, second, that the *solution* to the epistemological problem is *technological*: more and better techniques and technologies, which will 'shrink' big data back to a manageable size. The epistemological problem is different, and it requires an epistemological solution.

Consider the problem first. 'Big data' came to be formulated after other buzz expressions, such as 'infoglut' or 'information overload', began to fade away, yet the idea remains the same. It refers to an overwhelming sense that we have bitten off more than we can chew, that we are being force-fed like geese, that our intellectual livers are exploding. This is a mistake. Yes, we have seen that there is an obvious exponential growth of data on an ever-larger number of topics, but complaining about such over-abundance would be like complaining about a banquet that offers more than we can ever eat. Data remain an asset, a resource to exploit. Nobody is forcing us to digest every available byte. We are becoming data-richer by the day; this cannot be the fundamental problem.

Since the problem is not the increasing wealth of data that is becoming available, clearly the solution needs to be reconsidered: it cannot be merely how many data we can technologically process. We saw that, if anything, more and better techniques and technologies are only going to generate more data. If the problem were too many data, more ICTs would only exacerbate it. Growing bigger digestive systems, as it were, is not the way forward.

The real epistemological problem with big data is *small patterns*. Precisely because so many data can now be generated and processed so quickly, so cheaply, and on virtually anything, the pressure both on the data *nouveau riche*, such as Facebook or Walmart, Amazon or Google, and on the data *old money*, such as genetics or medicine, experimental physics or neuroscience, is to be able to spot where the new patterns with real added-value lie in their immense databases, and how they can best be exploited for the creation of wealth, the improvement of human lives, and the advancement of knowledge. This is a problem of brainpower rather than computational power.

Small patterns matter because, in hyperhistory, they represent the new frontier of innovation and competition, from science to business, from governance to social policies, from security to safety. In a free and open marketplace of ideas, if someone else can exploit the small patterns earlier and more successfully than you do, you might quickly be out of business, miss a fundamental discovery and the corresponding Nobel, or put your country in serious danger.

Small patterns may also be risky, because they push the limit of what events or behaviours are predictable, and therefore may be anticipated. This is an ethical problem. Target, an American retailing company, relies on the analysis of the purchasing patterns of 25 products in order to assign each shopper a 'pregnancy prediction' score, estimate her due date, and send coupons timed to specific stages of her pregnancy. In a notorious case,[8] it caused some serious problems when it sent coupons to a family in which the teenager daughter had not informed her parents about her new status. I shall return to this sort of problem in Chapters 3 and 4, when discussing personal identity and privacy.

Unfortunately, small patterns may be significant only if properly aggregated, correlated, and integrated—for example in terms of loyalty cards and shopping suggestions—compared, as when banks utilize big data to fight fraudsters, and processed in a timely manner, as in financial markets. And because information is indicative also when it is not there (the lack of some data may also be informative in itself), small patterns can also be significant if they are absent. Sherlock Holmes solves one of his famous cases because of the silence of the dog, which should have barked. If big data are not 'barking' when they should, something is going on, as the financial watchdogs (should) know, for example.

Big data is here to grow. The only way of tackling it is to know what you are or may be looking for. We do not do science by mere accumulation of data; we should not do business and politics in that way either. At the moment, the required epistemological skills are taught and applied by a black art called analytics. Not exactly your standard degree at university. Yet so much of our well-being depends on it that it might be time to develop a methodological investigation of it. Who knows, philosophers might not only have something to learn, but also a couple of lessons to teach. Plato would agree. What he might have been disappointed about is the fact that memory is no longer an option. As we shall see in Chapter 7, memory may outperform intelligence, but mere data hoarding, while waiting for more powerful computers, smarter software, and new human skills, will not work, not least because we simply do not have enough storage. Recall our future historian: this is the first bottleneck she identified in the development of hyperhistory, which suffers from digital amnesia.

Memory

Hyperhistory depends on big data, but there are two myths about the dependability of digital memory that should be exposed in this first chapter.

The first myth concerns the *quality* of digital memory. ICTs have a kind of forgetful memory. They become quickly obsolete, they are volatile, and they are rerecordable. Old digital documents may no longer be usable because the corresponding technology, for example floppy drives or old processing software, is no longer available. There are millions of abandoned pages on the Internet, pages that have been created and then not updated or modified. At the beginning of 1998, the average life of a document that had not been abandoned was 75 days. It is now estimated to be 45 days. The outcome is that so-called link decay (links to resources online that no longer work) is a common experience. On 30 April 1993, the European Organization for Nuclear Research (CERN) announced that the World Wide Web it had created would be free to everyone, with no fees due. Twenty years later, to celebrate the event, a team at CERN had to recreate the first web page (with its original URL etc.), because it no longer existed. Our digital memory seems as volatile as our oral culture was but perhaps even more unstable, because it gives us the opposite impression. This paradox of a digital 'prehistory'—ICTs are not preserving the past for future consumption because they make us live in a perennial present—will become increasingly pressing in the near future. Memory is not just a question of storage and efficient management; it is also a matter of careful curation of significant differences, and hence of the stable sedimentation of the past as an ordered series of changes, two historical processes that are now seriously at risk. Ted Nelson, for example, a pioneer in ICTs who coined the terms 'hypertext' and 'hypermedia', designed Xanadu so that it would never delete copies of old files. A website constantly upgraded is a site without memory of its own past, and the same dynamic system that allows one to rewrite a document a thousand times also makes it unlikely that any memory of past versions will survive for future inspection. 'Save this document' means 'replace its old version', and every digital document of any kind may aspire to such an ahistorical nature. The risk is that differences are erased, alternatives amalgamated, the past constantly rewritten, and history reduced to the perennial here and now. When most of our knowledge is in the hands of this forgetful memory, we may find ourselves imprisoned in a perpetual present. This is why initiatives aimed at preserving our increasingly digital cultural heritage for future generations—such as the National Digital Stewardship Alliance (NDSA) and the International Internet Preservation Consortium (IIPC)—are vital. The job of information curators is bound to become ever more important.

There is then the potentially catastrophic risk of immense quantities of data being created simultaneously. We saw that most, indeed almost all, of our data have been created in a matter of a few years. They are all getting old together, and will reach the threshold of system failure together, like a baby-boom generation retiring at the same time. To understand the problem, recall the old debate about your collection of music CDs and how they would all be unusable within a decade, as opposed to your vinyl records. According to the Optical Storage Technology Association, the shelf life of new, unrecorded CDs and DVDs is conservatively estimated to be between 5 and 10 years. And according to the National Archives and Records Administration,[9] once recorded, CDs and DVDs have a life expectancy of 2 to 5 years, despite the fact that published life expectancies are often cited as 10 years, 25 years, or longer. The problem is that after a few years the material degrades too much to guarantee usability. The same applies to our current digital supports, hard disks and memories of various kinds. The 'mean time before failure' (MTBF) figure indicates an estimate of a system's life expectancy.[10] The higher the MTBF, the longer the system should last. An MTFB of 50,000 hours (5.7 years) for a standard hard disk is not uncommon. This short life expectancy is already a problem. But the real issue that I am stressing here is another. Contrary to what we experienced in the past, the life expectancies of our data supports are today dangerously synchronized. This is why you may think of this as a sort of 'baby boom': big data will age and become dead data together. Clearly, huge quantities of data will need to be rerecorded and transferred to new supports at regular intervals. Indeed they already are. But which data are going to make it to the other side of any technological transition? For a comparison, consider the transition of silent movies to new kinds of support, or of recorded music from vinyl to the CD. Huge quantities of data were left behind, becoming lost, unavailable, or inaccessible.

According to a 2012 Research Report by IBIS World, the data recovery industry saw its overall revenue over the five years to 2012 fall at an annualized rate of 0.9 per cent to total $1 billion, with a decline of 0.6 per cent in 2012.[11] This may seem counter-intuitive. Big data is growing and so are the problems concerning damaged, corrupted, or inaccessible files and storage media. The industry that takes care of such problems should be flourishing. The explanation is that cloud or online storage has expanded the options for data recovery and data loss prevention. If you use Dropbox, Google Docs, Apple iCloud, or Microsoft Skydrive, for example, and your computer is damaged, the files are still available online and can be easily recovered, so you will not need a data recovery service. Yet, this seems to be just a question of transition and hence time. Cloud computing has put pressure on an industry specialized in computers at a consumer level. The more our gadgets become mere terminals, the less we need to worry ourselves about the data. But the storage of those data still relies on physical infrastructures, and these will need increasing maintenance. The data recovery industry will disappear, but a new industry dedicated to cloud computing failures is already emerging. It is not a matter of relying on the brute force of redundancy (having more than one copy of the same file). This strategy is not available at a global level, because of the second myth about the dependability of digital memory, the one concerning the *quantity* of digital memory.

Since 2007, the world has been producing many more data than available storage.[12] This despite the fact that, according to Kryder's Law (another generalization), storage density of hard disks is increasing more quickly than Moore's Law, so that it is predicted that in 2020 a disk of 14 terabytes will be 2.5 inches in size and will cost about $40. Unfortunately, this will not be enough, because even the growth projected by Kryder's Law is slow when compared to the pace at which we generate new data. Think of your smartphone becoming too full because you took too many pictures, and make it a global problem. In history, the problem was what to save: which laws or names were going to be baked in clay or carved in stone, which texts were going to be handwritten on papyrus or vellum, which news items were worth printing on paper. In hyperhistory, saving is the default option. The problem becomes what to erase. Since storage is insufficient, something must be deleted, rewritten, or never be recorded in the first place. By default the new tends to push out the old, or 'first in first out': updated web pages erase old ones, new pictures make old ones look dispensable, new messages are recorded over old ones, recent emails are kept at the expense of last year's.

Hyperhistory ran out of memory space in which to dump its data many years ago. There is no name for this 'law' about the increasing shortage of memory, but it looks like the gap is doubling every year. Barring some significant technological breakthrough in physical storage or software compression, the process will get worse, *quantitatively*. The good news is that it does not have to be as bad as it looks, *qualitatively*. Rephrasing a common saying in the advertising industry, half of our data is junk, we just do not know which half. You are happy to take ten pictures because you hope one will come out right, and the other nine can be discarded. They were never intended to be saved in the first place. This means that we need a much better understanding of which data are worth preserving and curating. This, in turn, is a matter of grasping which questions are, or will be, interesting to ask not only now, but also in the future, as we saw in the previous section. And this leads to a slightly reassuring virtuous circle: we should soon be able to ask big data what data are worth saving. Think of an app in your smartphone not only suggesting which of the ten pictures is worth keeping, but also learning from you, once you have taken a decision (maybe you prefer darker pictures). Then new challenges will concern how we may avoid poor machine-based decisions, improve so-called 'machine learning', or indeed make sure machines relearn new preferences (later in life you may actually like brighter pictures). More information may help us to decide which information to save and curate. Our future historian may well interpret the zettabyte age of hyperhistory as the time of transition between blind and foresighted big data.

So much for the first bottleneck: shortage of memory. To understand the other, concerning connectivity, we need to look first at some features of networks.

Connectivity

Computers may be of limited use if they are not connected to other computers. This was not always obvious. Sometimes it is still questionable, as when your computers need to be hacker-proof because they control the launch of nuclear missiles, for example. But, in general, the observation is rather trivial today. In the age in which tethering no longer means restraining an animal with a rope or chain (the tether) but actually connecting one ICT device to another, the question is no longer whether connectivity has any value, but how much value it actually has. Many theories and laws have been proposed: Reed's Law, Sarnoff's Law, Beckstrom's Law... but the most famous remains Metcalfe's. Like the laws just mentioned and Moore's Law, it is a generalization ('this is how things tend to go, more or less') rather than a scientific law, but it is, nevertheless, enlightening. It states that the value of a network is proportional to the square of the number of connected nodes of the system (n^2). So a network of two computers has only a value of $2^2=4$, but doubling the number of connected computers already means quadrupling the value of their network to $4^2=16$. Figure 11 illustrates what happens after 20 iterations. The idea is simple: the more nodes you have, the more useful it is to be connected and expensive to be unconnected. Indeed, the point to keep in mind is even simpler, for there is an even more inclusive generalization. Any growth bigger than linear (a linear growth is when you multiply x by a fixed number, like your salary by 12 months), e.g. squared, like Metcalfe's, or cubic (n^3), or exponential (e^x), after a few iterations looks like a straight perpendicular line, like a capital L which has been rotated 180 degrees on its axis:

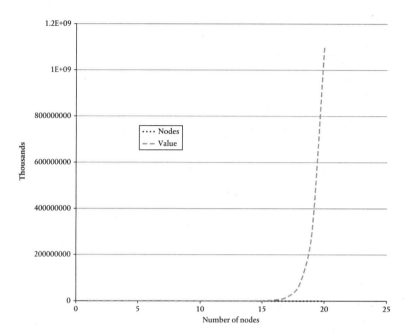

Fig. 11. Metcalfe's Law: The Value of a Network of n Nodes=n^2.

This 'L law' is all one needs to remember. It is the shape of growth that any business would like to achieve. It is the shape of hyper-connectivity. According to a report by the International Telecommunication Union (ITU),[13] in 2013 more than a third of the world population was online. No wonder the value of the network has skyrocketed, straight like a rod. So what is the problem? Any L Law does not really address the communication within the network, but rather the value of its complexity (how many links are possible among how many nodes). Communication requires a link but it comes with a speed. Think of a road, and the difference it makes whether it is a small street or a motorway, with or without traffic. This is the bottleneck our future historian identified. It is known as Nielsen's Law.

Some years ago, Jacob Nielsen noticed that, in general, the speed of network connections for home users like you and me increases approximately 50 per cent per year, thus doubling every 21 months or so. This is impressive, but not as impressive as the speed identified by Moore's Law. It is also already insufficient to cope with the faster growing 'weight' (number of bits) of the files we wish to transfer. As a result, for the foreseeable future our online experience will be constrained by our bandwidth.

Conclusion

The living generation is experiencing a transition from history to hyperhistory. Advanced information societies are more and more heavily dependent on ICTs for their normal functioning and growth. Processing power will increase, while becoming cheaper. The amount of data will reach unthinkable quantities. And the value of our network will grow almost vertically. However, our storage capacity (space) and the speed of our communications (time) are lagging behind. Hyperhistory is a new era in human development, but it does not transcend the spatio-temporal constraints that have always regulated our life on this planet. The question to be addressed next is: given all the variables we have seen in this chapter, what sort of hyperhistorical environment are we building for ourselves and for future generations? The short answer is: the *infosphere*. The long answer is provided by Chapter 2.

2

SPACE

Infosphere

Technology's in-betweenness

One of the most obvious features that characterizes any technology is its *in-betweenness*. Suppose Alice lives in Rio de Janeiro, not in Oxford. A hat is a technology between her and the sunshine. A pair of sandals is a technology between her and the hot sand of the beach on which she is walking. And a pair of sunglasses is between her and the bright light that surrounds her. The idea of such an in-betweenness seems clear and uncontroversial. However, it soon gets complicated.

Because of our anthropocentric concerns, we have a standard term to describe one of the sides of technology's in-betweenness: Alice is the *interacting user*. What we seem to lack is a term for the other side of the relation, that which invites a particular usage or enables some interaction. What the sun does is to prompt the development and then the wearing of the hat. So let us agree to refer to the other side of technology's in-betweenness as *the prompter*.[1] Apart from conveying the right idea of inviting, suggesting, or enabling some particular technological mediation, it is also a virgin word in our philosophy of technology, hard to confuse with its meaning in the theatre, and it rhymes with user. Here (see Figure 12), it means that the sunshine is a prompter of the hat, the hot sand is a prompter of the sandals, and the bright light is a prompter of the sunglasses. An inventor is someone

Fig. 12. The Scheme for Technology's In-betweenness.

Fig. 13. First-order Technology.

who devises an artefact that may satisfy a user's need or want caused by some prompter. As you can see, I am slightly stretching the word 'prompter', hopefully without breaking it.

When technologies are in-between *human* users and *natural* prompters, we may qualify them as *first-order* (Figure 13). Listing first-order technologies is simple. The ones mentioned earlier all qualify. More can easily be added, such as the plough, the wheel, or the umbrella. The axe is probably the first and oldest kind of first-order technology. Nowadays, a wood-splitting axe is still a first-order technology between you, the user, and the wood, the prompter. A saddle is between you and a horse. Nail clippers and hunting bows are other instances of such first-order kind of technology, which need not be simple, and can be technology-dependent and technically sophisticated, like an assault rifle, which is sadly a first-order technology between two human sides, as users and prompters.

At this point, the word 'tool' may come to mind as appropriate, but it would be misrepresentative, because tools do not have to be first-order technologies, as I shall explain presently.

Many non-human animals make and use simple, first-order technologies, like modified sticks or shells, to perform tasks such as foraging, grooming, fighting, and even playing. In the past, this discovery determined the end of a naïve interpretation of *homo faber* as *homo technologicus*. True, we are the species that builds, but the point to be made is slightly subtler, because many other species also create and use artefacts to interact with their environments. As in the case of our

Fig. 14. Second-order Technology.

use of natural languages and other symbolic forms of communication, or the creation of artificial languages, especially to program machines, the difference between us and other species is incommensurable not because it is a matter of binary presence or absence of some basic abilities, but because of the immensely more sophisticated and flexible degree to which such abilities are present in us. It is the difference between a colouring book with which a child has played using some crayons and the Sistine Chapel. Insisting on continuity is not mistaken, it is misleading. In the case of technologies, it is preferable to talk about *homo faber* as *homo technologicus*, inventor and user of *second-* and *third-order* technologies, in the following sense.

Second-order technologies are those relating users no longer to nature but to other technologies; that is, they are technologies whose prompters are other technologies (see Figure 14).

This is a good reason not to consider the concept of a *tool* or that of a *consumer good* as being coextensive with that of first-order technology. Think of the homely example of a humble screwdriver. Of course, it is a tool, but it is between you and, you guessed it, a screw, which is actually another piece of technology, which in its turn (pun irresistible) is between the screwdriver and, for example, two pieces of wood. The same screwdriver is also more easily understood as an instance of a *capital good*, that is, a good that helps to produce other goods. Other examples of such second-order technologies include keys, whose prompters are obviously locks, and vehicles whose users are (still) human and whose prompters are paved roads, another piece of technology.

Some first-order technologies (recall: these are the ones that satisfy the scheme humanity–technology–nature) are useless without the corresponding second-order technologies to which they are coupled.

Roads do not require cars to be useful, but screws call for screwdrivers. And second-order technologies imply a level of mutual dependency with first-order technologies (a drill is useless without the drill bits) that is the hallmark of some degree of specialization, and hence of organization. You either have nuts *and* bolts or neither.

Such interdependencies, and the corresponding appearance of second-order technologies, require trade and some kind of currency, so they are usually associated with the emergence of more complex forms of human socialization, and therefore some kind of civilization, the following accumulation of some free time and leisure, and ultimately a corresponding culture. Whereas some non-human animals are able to build their own artefacts to some extent, for instance by sharpening a stick, they do not seem to be able to build second-order technologies in any significant way.

The engine, understood as any technology that provides energy to other technologies, is probably the most important second-order technology. Watermills and windmills converted energy into useful motion for millennia, but it is only when the steam, the internal combustion engine, and the electric motor become 'portable' energy-providers, which can be placed between users and other technologies wherever they are needed, that the Industrial Revolution turns into a widespread reality.

Much of late modernity—prompted by science's increasing knowledge about, and control over, materials and energy—gets its mechanical aftertaste from the preponderance of second-order technologies. The London of Sherlock Holmes is a noisy world of gears, clocks, shafts, wheels, and powered mechanisms, characterized not just by the humanity–technology–nature relation but, more significantly, by the humanity–technology–technology relation. Modernity, as a pre-hyperhistorical stage of human development, soon becomes a world of complex and networked dependencies, of mechanical chain-reactions as well as locked-in connections: no trains without railways and coal, no cars without petrol stations and oil, and so forth, in a mutually reinforcing cycle that is both robust and constraining.

As the history of the floppy disk shows, at some stage it is easier to replace the whole system—change paradigm, to put it more dramatically—than to keep improving one part of it. There is no point in having super-powerful floppy disks if the millions of drives already in place are not up to the task of reading them. This explains one of the advantages of any technological leapfrogging:[2] a later adopter does not have to deal with the legacy of any incumbent technological package (coupled first- and second-order technology), and is free to take advantage of the most recent and innovative solution. Yet, this is less simple than it looks, precisely because of the coupled nature of second-order technologies. Of course, it would be easier to introduce electric or hybrid vehicles, for example, if there were only roads but no internal combustion engine vehicles; the obvious difficulty is that roads are there because of the latter in the first place. Thus, the task of legislation that deals with technological innovation is also that of easing the transition from old to new technologies by decoupling, sometimes through incentives and disincentives, what needs to be kept (e.g., roads) from what needs to be changed (e.g., internal combustion engine vehicles).

Most of the comfortable appliances we enjoy in our houses today belong to late modernity, in terms of conception: the refrigerator, the dishwasher, the washing machine, the clothes dryer, the TV, the telephone, the vacuum cleaner, the electric iron, the sound system…these are all either first- or second-order technologies, working between human users and the relevant prompters. They represent a world that is ripe for a third-order, revolutionary leap. For technology starts developing exponentially once its in-betweenness relates technologies-as-users to other technologies-as-prompters, in a technology–technology–technology scheme (see Figure 15). Then we,

Fig. 15. Third-order Technology.

who were the users, are no longer in the loop, but at most on the loop: *pilots* still fly drones actively, with a stick and a throttle, but *operators* merely control them with a mouse and a keyboard.[3] Or perhaps we are not significantly present at all, that is, we are out of the loop entirely, and enjoy or simply rely on such technologies as (possibly unaware) beneficiaries or consumers. It is not an entirely unprecedented phenomenon. Aristotle argued that slaves were 'living tools' for action:

> An article of property is a tool for the purpose of life, and property generally is a collection of tools, and a slave is a live article of property.[4] [...] These considerations therefore make clear the nature of the slave and his essential quality; one who is a human being belonging by nature not to himself but to another is by nature a slave, and a human being belongs to another if, although a human being, he is a piece of property, and a piece of property is an instrument for action separate from its owner.[5]

Clearly, such 'living tools' could be 'used' as third-order technology and place the masters off the loop. Today, this is a view that resonates with many metaphors about robots and other ICT devices as slaves.

Of course, the only safe prediction about forecasting the future is that it is easy to get it wrong. Who would have thought that, twenty years after the flop of Apple's Newton[6] people would have been queuing to buy an iPad? Sometimes, you just have to wait for the right apple to fall on your head. Still, 'the Internet of things', in which third-order technologies work independently of human users, seems a low-hanging fruit sufficiently ripe to be worth monitoring. Some pundits have been talking about it for a while now. The next revolution will not be the vertical development of some unchartered new technology, but horizontal. For it will be about connecting anything to anything (*a2a*), not just humans to humans. One day, you-name-it 2.0 will be *passé*, and we might be thrilled by *a2a* technologies. I shall return to this point in Chapter 7. For the moment, the fact that the Newton was advertised as being able to connect to a printer was quite amazing at the time, but rather trivial today. Imagine a world in which your car autonomously checks your electronic diary and reminds you, through your digital TV, that you need to get some petrol tomorrow, before your long-distance commuting. All this and more is already easily feasible. The greatest obstacles are a lack of shared standards, limited protocols, and hardware that is not designed to be fully modular with the rest of the infosphere. Anyone who could invent an affordable, universal appliance that may be attached to our billions of artefacts in order to make them interact with each other would soon be a billionaire. It is a problem of integration and defragmentation, which we currently solve by routinely forcing humans to work like interfaces. We operate the fuel dispensers at petrol stations, we translate the GPS's instructions into driving manoeuvres, and we make the grocery supermarket interact with our refrigerator.

Essentially, third-order technologies (including the Internet of things) are about removing us, the cumbersome human in-betwe-eners, off the loop. In a defragmented and fully integrated infosphere, the invisible coordination between devices will be as seamless as the way in which your smartphone interacts with your laptop and the latter interacts with the printer. It is hard to forecast what will happen when things regularly talk to each other, but I would not be surprised if computer and software companies will design and sell appliances, including your TV, in the near future.

Technologies as users interacting with other technologies as prompters, through other in-between technologies: this is another way of describing hyperhistory as the stage of human development when third-order technological relations become the necessary con-dition for development, innovation, and welfare. It is also a way of providing further evidence that we have entered into such a hyperhis-torical stage of our development. The very expression 'machine-readable data' betrays the presence of such a generation of third-order technologies. To put it simply, barcodes are not for our eyes, and in high-frequency trading[7] (three-quarters of all equity trading volume in the US is HFT) the buying and selling of stocks happens at such an extremely high speed that only fast computers and algorithms can cope with it, scanning many marketplaces simultaneously, executing millions of orders a second, and adopting and adapting strategies in milliseconds. The same holds true in any time-sensitive application, whether civilian or military. Further examples include autonomous vehicles, like driverless cars, or 'domotic appliances', the technologies that are transforming the house into a smart environment, for instance by monitoring, regulating, and fine-tuning the central heating and the supply of hot water to our habits. We shall encounter them again in the following chapters.

As is clear from the previous examples, the ultimate third-order technology is provided by ICTs. The very use of 'engine' in computational contexts (as in 'search engine' or 'game development engine') reminds us that second-order technology is related to the engine as third-order technology is related to the computer. ICTs can process data autonomously and in smart ways, and so be in charge of their own behaviours. Once this feature is fully exploited, the human user may become redundant. It is hard to imagine a modern world of mechanical engines that keeps working and repairing itself once the last human has left Earth. Mechanical modernity is still human-dependent. However, we can already conceive a fully automated, computational system that may not need human interactions at all in order to exist and grow. Projects to build self-assembling 3D printers that could exploit lunar resources to build an artificial colony on the Moon may still sound fictional,[8] but they illustrate well what the future looks like. Smart and autonomous agents no longer need to be human. A hyperhistorical society fully dependent on third-order technologies can in principle be human-independent.

Time to summarize. We saw that technologies can be analysed depending on their first-, second-, or third-order nature. The point could be refined, but without much conceptual gain. Is a clock a first- (between you and your time), a second- (between you and your pressure cooker), or a third-order technology (between your computer and some scheduled task)? Is a pair of scissors a first- (between you and the stem of a rose), a second- (between you and a piece of paper), or a third- (between a robot and a piece of cloth in a factory) order technology? Is a computer a first- (between you and the level of water in a reservoir), a second- (between you and another computer), or a third-order (between two other computers) technology? Evidently each answer depends on the context. Yet, the fact that there is no single, decontextualized answer does not make the distinction any less cogent, it only proves that we need to be careful when using it. What is important to stress here is that the distinction is both sound and complete: there is no fourth-order technology. Of course, the chain of technologies interacting with other technologies can be extended as much as one wishes. However, such a chain can always be reduced to a series of triples, each of which will be of first-, second-, or third-order.[9]

The development of technologies, from first- to second- and finally to third-order, poses many questions. Two seem to be most relevant in the context of our current explorations.

First, if technology is always in-between, what are the new related elements when ICTs work as third-order technologies? To be more precise, for the first time in our development, we have technologies that can regularly and normally act as autonomous users of other technologies, yet what is ICTs' in-between relationship to us, no longer as users but as potential beneficiaries who are out of the loop? A full answer must wait until the following chapters. Here, let me anticipate that, precisely because ICTs finally close the loop, and let technology interact with technology through itself, one may object that the very question becomes pointless. With the appearance of third-order technologies all the in-betweenness becomes internal to the technologies, no longer our business. We shall see that such a process of technological 'internalization' has raised concern that ICTs may end up shaping or even controlling human life. At the same time, one may still reply that ICTs, as third-order technologies that close the loop, internalize the technological in-betweenness but generate a new 'outside', for they create a new space (think for example of cyber-space), which is made possible by the loop, that relies on the loop to continue to exist and to flourish, but that is not to be confused with the space inside the loop. Occurrences of such spaces are not socially unprecedented. At different times and in different societies, buildings have been designed with areas to be used only by slaves or servants for the proper, invisible functioning of the whole house-system, from the kitchen and the canteen to separate stairs and corridors. Viewers of the popular TV drama *Downton Abbey* will recognize the 'upstairs, down-stairs' scenario. What is unprecedented is the immense scale and pace at which the whole of human society is now migrating to this out-of-the-loop space, whenever possible.

Second, if technology is always in-between, then what enables such in-betweenness to be successful? To put it slightly differently: how does technology interact with the user and the prompter? The one-word answer is: interfaces, the topic of the next section.

Interfaces

Janus is the Roman god of passages and transitions, endings and beginnings, both in space (like thresholds, gates, doors, or borders) and in time (especially the end of the old and the beginning of the new year, hence January, or of different seasons, or of times of peace and war, etc.). Janus is easily recognizable among the gods, because he is represented as having two faces (bifront). Nowadays, he is our god of interfaces and presides over all digital technologies, which are by definition bifront.

One face of our bifront ICTs looks at the user and it is expected to be friendly. The other face connects the technology in question to its prompter. We may call it protocol, although, strictly speaking, this term is used only to refer to the set of rules that regulate data transmission. Any order of technology has two faces, the user inter-face and the protocol. Think of these two faces as being represented by the two connectors in the 'user–technology–prompter' scheme (see Figure 16).

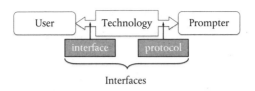

Fig. 16. Technology's Interfaces.

Depending on the order of the technological in-betweenness, the protocol face may become increasingly less visible, with the user's interface ending up being treated as *the* interface, until it too may disappear. Janus starts hiding his other side and looking more like any other god, single-faced, until even that face is no longer perceivable. Let me explain this gradual disappearance by using the examples introduced before.

We saw that the wood-splitting axe is a typical first-order technology. It fits the scheme humanity–technology–nature. The grip is the user-friendly interface, and the handle+blade is the protocol that connects the interface (transmits the force of the blow) to the prompter, the wood. Because you, the user, need to control both your interaction with the technology in-between and its interaction with the natural prompter, you have access to both of Janus' faces, the user's interface and the protocol. You can easily sharpen the blade, for example.

Consider next the quintessential second-order technology, the engine. We now have a case of the humanity–technology–technology scheme. Technological protocols may now ensure that the technology in-between takes care of the technological prompter. In some cases, you still need access to the protocol. Think of the screwdriver, and how you need to see whether the tip of its blade corresponds to, and fits, the head of the screw (are they both slotted, Phillips, Robertson etc.?). But usually, you, still the user, do not have to have access to both faces. All you perceive and interact with, for example, is the gear stick and the clutch pedal in a manual-transmission car. The protocol—that is, how the other face of the engine interacts with the prompter represented by the rest of the car's propelling system—is not important, as long as there is no malfunctioning. For this reason, in second-order technologies we no longer tend to distinguish between the interface in general (which includes also the protocol) and the user's interface. At this point, 'interface' simply refers only to the user's interface, since the protocol is no longer quite so obvious or accessible. If something goes wrong, access to the protocol and to the prompter often requires a specialist.

Finally, consider a generic modem as a case of third-order technology. As the word indicates, this is a device that *mo*-dulates an analogue signal to encode digital information at the sender's side, and *dem*-odulates such a signal to decode the transmitted information at the receiver's side, often over a telephone line. We now have a case of technology–technology–technology scheme (if you find the modem too passé, consider a router). Since the interface (the connectors in the scheme) now connects technology to technology through some other technology, and third-order technology requires autonomous processing capacities, made possible by ICTs, the tendency is to interpret the whole interface as a set of protocols. Technological protocols ensure that the technology in-between, the two modems, take care both of the technological user, let's say your computer, and of the technological prompter, let's say my computer. The dynamic, automated process of negotiation of protocols, which sets the necessary and sufficient parameters required for any further communication, is known as *handshaking*. It is what goes on between your and my computer, between your computer and your printer, when they 'see' each other, or between your smartphone and your laptop, before they can agree to synchronize your digital diary. You and I are neither invited to, nor involved in, such a handshaking. Now both faces of Janus may be hidden from us. We are out of the loop entirely. You go home and your smartphone automatically connects to your home wireless service, downloads some updates, and starts 'talking' to other ICT devices in the house, like your tablet. As in any 'plug and play' case, all the required handshaking and the issuing data processing is invisible to the ultimate beneficiaries, us. As in a classic Renaissance house, we now inhabit the *piano nobile*, the upper, noble floor, not even knowing what happens in the ground floor below us, where technologies are humming in the service rooms. Unless there is some malfunctioning, we may not even know that such technologies are in place. But, if something goes wrong, it is the specialist who will now have to take care of both sides of the interface, with the result that specialists are the new priests in Janus' temple. They will become increasingly powerful and influential the more we rely on higher-order technologies.

Design

Interfaces, like the technologies with which they are associated, evolve. Such an evolution is made possible by, among many things, design. It is often a successful story of improvements, even when the technology in question may be distasteful, at least to a pacifist. If you visit a military museum and look at very old, handheld gunpowder weapons, you will notice that it took a surprisingly long time before manufacturers developed what seems now utterly obvious: a hand-friendly grip. So-called hand cannons,[10] which originated in China and became popular in Europe during the Renaissance, consisted of barrels that would be aimed at enemies more like crossbows. For a long time, old handguns were almost as straight as swords.[11] They begun to bend only slowly through time and acquired the familiar L-shape quite late. This is surprising, given how obvious this design now looks to any kid who points a finger gun at you, his thumb raised above his fist, with one or two straight fingers acting as a barrel.

Sometimes, design can be intentionally retro. Apple's iMac G3, the first model of the iMac series, was, like its successors, an all-in-one computer, including in a single enclosure both the monitor and the system unit. Among its peculiarities was the fact that the casing was made of colourful, translucent plastic. You had the impression of being able to see the inside of the machine, that is, both faces of

Janus, your user's interface and the protocols on the other side. Thus, it had a friendly, first-order technology look—recall the axe—when in fact it could be used as a sophisticated, second-order system, in-between the human user and some other technological artefacts. As in the case of the engine of your car, you did not need to see the inside, and actually you did not really see the protocols, and could do nothing about them even if you saw them. It was modern aesthetics at play, functionally pointless. It did not last.

Sometimes, the design may be simply outdated, a legacy from the past. Front-loading washing machines developed from mechanical laundry systems. So they still have a door with a transparent window to check whether there is water inside—this is one of the most plausible explanations provided when they are compared to top-loading ones, which lack transparent windows—even if, in fact, you can no longer open the door if this is unsafe. The later-developed dishwasher never had a see-through door like a washing machine.

The design of good interfaces takes time and ingenuity. It may be a matter of realizing the obvious (a hand-friendly handle) or removing the pointless (a no longer useful transparent window). You do not need a bright light to signal that your computer is switched on; so many computers do not have one. But these days you still need easy access to a USB port for your USB flash drive, and having the port in the back of the computer, as some computers do, may be visually elegant but functionally cumbersome. The conclusion is that, in terms of interfaces' functionality (and there are of course other terms, including usability, economic, aesthetic, ergonomic, or energy-related ones, for example), good design is design that takes into account and makes the most of the ordering of the technology in question. In first-order technologies, both the user's interface and the protocol need to be accessible and friendly. In second-order technologies, good design needs to concentrate only on the user-friendly face of the interface, while the protocol can be invisible. It is pointless to have a transparent case for a watch that is not even meant to be repairable. In third-order technologies, both sides of the interface, the user's and the protocol, should become functionally invisible to us. Yet such functional invisibility contributes to making the question about the in-betweenness of third-order technologies more pressing. Being out of the loop could mean being out of control. Such a concern soon turns into a political issue, as we shall see in Chapter 8. Here, let me just draw a caricature in black and white, without any nuances, for the sake of a quick and simple clarification.

The politics of technology

Interpretations of the politics of technology's in-betweenness—in more prosaic terms, the dynamics of technological R & D, deployments, uses, and innovations, all more or less shaped by human aggregate decisions, choices, preferences, mere inertia, and so forth— may swing between two extremes. No serious scholar advocates either of them, but they help to convey the basic idea.

At one extreme, one may interpret technology's in-betweenness as a deleterious kind of detachment and a loss of pristine contact with the natural and the authentic. This position may go as far as to associate technology's in-betweenness with disembodiment or at least a devaluation of embodiment, hence to delocalization (no body, no place), globalization (no place, no localization), and ultimately with consumerism, as a devaluation of the uniqueness of physical things and their special relations with humans. In this case, the politics of technological in-betweenness assume the features of, at best, a lamentable, global mistake, and, at worst, of an evil plan, single-mindedly pursued by some malevolent agents, from states to corporate multinationals.

At the other extreme, there is the enthusiastic and optimistic support for the liberation provided by technology's in-betweenness. This is interpreted as a buffer, as a way of creating more space for communication and personal fulfilment. The idea of technological in-betweenness is not seen as a dangerous path towards the exercise of power by some people, systems, or even machines over humans, but as an empowering and enabling form of control. The equation may run somewhat like this: more space = more freedom = more control = more choice.

Clearly, neither extreme position is worth taking seriously. However, various combinations of these two simple ingredients dominate our current discussion of the politics of technology. We shall see in the following chapters that debates soon become messier and much less clear cut.

ICTs as interpreting and creating technologies

Today, when we think of technology in general, ICTs and their ubiquitous, user-friendly interfaces come immediately to mind. This is to be expected. In hyperhistorical societies, ICTs become the characterizing first-, second-, and third-order technologies. We increasingly interact with the world and with our technologies through ICTs, and ICTs are *the* technologies that can, and tend to, interact with themselves, and invisibly so. Also to be expected is that, as in the past, the dominant technology of our time is having a twofold effect. On the one hand, by shaping and influencing our interactions with the world, first- and second-order ICTs invite us to interpret the world in ICT-friendly terms, that is, informationally. On the other hand, by creating entirely new environments, which we then inhabit (the out-of-the-loop experience, functionally invisible by design), third-order ICTs invite us to consider the intrinsic nature of increasing portions of our world as being inherently informational. In short, ICTs make us think about the world informationally and make the world we experience informational. The result of these two tendencies is that ICTs are leading our culture to conceptualize the whole reality and our lives within it in ICT-friendly terms, that is, informationally, as I shall explain in this section.

ICTs are modifying the very nature of, and hence what we mean by, reality, by transforming it into an infosphere. Infosphere is a neologism coined in the seventies. It is based on 'biosphere', a term referring to that limited region on our planet that supports life. It is also a concept that is quickly evolving. *Minimally*, infosphere denotes the whole informational environment constituted by all informational entities, their properties, interactions, processes, and mutual relations. It is an environment comparable to, but different from, cyberspace, which is only one of its sub-regions, as it were, since the infosphere also includes offline and analogue spaces of information. *Maximally*, infosphere is a concept that can also be used as synonymous with reality, once we interpret the latter informationally. In this case, the suggestion is that what is real is informational and what is informational is real.[12] It is in this equivalence that lies the source of some of the most profound transformations and challenging problems that we will experience in the near future, as far as technology is concerned.

The most obvious way in which ICTs are transforming the world into an infosphere concerns the transition from analogue to digital and then the ever-increasing growth of the informational spaces within which we spend more and more of our time. Both phenomena are familiar and require no explanation, but a brief comment may not go amiss. This radical transformation is also due to the fundamental convergence between digital tools and digital resources. The intrinsic nature of the tools (software, algorithms, databases, communication channels and protocols, etc.) is now the same as, and therefore fully compatible with, the intrinsic nature of their resources, the raw data being manipulated. Metaphorically, it is a bit like having pumps and pipes made of ice to channel water: it is all H_2O anyway. If you find this questionable, consider that, from a physical perspective, it would be impossible to distinguish between data and programs in the hard disk of your computer: they are all digits anyway.

Such a digital uniformity between data and programs was one of Turing's most consequential intuitions. In the infosphere, populated by entities and agents all equally informational, where there is no physical difference between *processors* and *processed*, interactions become equally informational. They all become interpretable as 'read/write' (i.e., access/alter) activities, with 'execute' the remaining type of process. If Alice speaks to Bob, that is a 'write' process, Bob listening to her is a 'read' process, and if they kiss, then that is an instance of 'execute'. Not very romantic, but accurate nonetheless.

Digits deal effortlessly and seamlessly with digits. This potentially eliminates one of the long-standing bottlenecks in the infosphere and, as a result, there is a gradual erasure of *informational friction*. I am only introducing this topic here, since Chapter 5 is entirely dedicated to it. For the moment, consider 'informational friction' as a label for how difficult it may be to let some information flow from sender to receiver. For example, in a noisy environment, like a pub or a cocktail party, you need to shout and maybe even use some gestures (that is, add redundancy) to ensure that your message gets across. If you wish to order two beers, you may use both your voice and some gestures. Because of their 'data superconductivity', ICTs are well known for being among the most influential factors that facilitate the flow of information in the infosphere. We are all acquainted daily with aspects of a *frictionless infosphere*, such as *spamming* (because every email flows virtually free) and *micrometering* (because every fraction of a penny may now count). Such 'data superconductivity' has at least four important consequences.

First, we are witnessing a substantial erosion of the *right to ignore*. In an increasingly frictionless infosphere, it becomes progressively less credible to claim one did not know when confronted by easily predictable events and hardly ignorable facts.

Second, there is an exponential increase in *common knowledge*. This is a technical term from logic, where it basically refers to cases in which everybody not only knows that p but also knows that everybody knows that everybody knows, …, that p. Think of a circle of friends sharing some information through a social media.

Third, the impact of the previous two phenomena is also quickly increasing because meta-information about how much information is, was, or should have been available is becoming overabundant. It follows that we are witnessing a steady increase in agents' *responsibilities*. The more any bit of information is just an easy click away, the less we shall be forgiven for not checking it. ICTs are making humanity increasingly responsible, morally speaking, for the way the world is, will be, and should be. This is a bit paradoxical since ICTs are also part of a wider phenomenon that is making the clear attribution of responsibility to specific individual agents more difficult and ambiguous.

The last consequence concerns informational privacy, but its analysis is too important so I shall delay it until Chapter 5.

Life in the infosphere

During the last decade or so, we have become accustomed to interpreting our life online as a mixture between an adaptation of human agents to digital environments (Internet as freedom from constraints and freedom of pursuits), and a form of postmodern, neo-colonization of digital environments by human agents (Internet as control). This is probably a mistake. We saw that ICTs are as much modifying our world as they are creating new realities and promoting an informational interpretation of every aspect of our world and our lives in it. With interfaces becoming progressively less visible, the threshold between *here* (*analogue, carbon-based, offline*) and *there* (*digital, silicon-based, online*) is fast becoming blurred, although this is as much to the advantage of the *there* as it is to the *here*. To adapt Horace's famous phrase, 'the captive infosphere is conquering its victor'.[13] The digital-online world is spilling over into the analogue-offline world and merging with it. This recent phenomenon is variously known as 'Ubiquitous Computing', 'Ambient Intelligence', 'The Internet of Things', or 'Web-augmented things'. I prefer to refer to it as the *onlife* experience. It is, or will soon be, the next stage in the development of the information age. We are increasingly living onlife.

The gradual informatization of artefacts and of whole (social) environments means that it is becoming difficult to understand what life was like in pre-digital times. In the near future, the distinction between online and offline will become ever more blurred and then disappear. For example, it already makes little sense to ask whether one is online or offline when driving a car following the instructions of a navigation system that is updating its database in real time. The same question will be incomprehensible to someone checking her email while travelling inside a driverless car guided by a GPS.

Sociologists speak of *Generation X*—people with birth dates from the early 1960s (that would be your author) to the early 1980s—and of *Generation Y*, or the Millennial Generation, which includes people with birthdates from the early 1980s to 2000 or so. Suppose then that we refer to people born after the long nineties—long because they lasted until 11 September 2001—as *Generation Z*, and not just because of the previous two alphabetical generations X and Y, but also because of the Zettabyte of data available to them. To people belonging to Generation Z, the world has always been wireless; 9/11 is a chapter in their elementary or primary school's history book; the Sistine Chapel has always been bright and colourful (restorations were unveiled in 1999). For them, there has never been a world without 'google', 'tweet', and 'wiki' not merely as services but as verbs; they have no recollection of a world without Facebook being a social media, not a book, and of books not being available online (Amazon was incorporated in 1994). They are likely to think that a pocket mirror is a phone app. They use Wikipedia (founded in 2001) as synonymous with encyclopaedia. To Generation Z or, more inclusively, to members of what Janna Quitney Anderson calls *Generation AO*, the Always-On Generation (see Figure 17), the peculiar clicking and whooshing sounds made by conventional modems while handshaking, also known as the whale song, are as archeologically alien as the sounds made by a telegraph's Morse signals to the ears of Generation X. Generation Z may not conceive of life outside the infosphere because, to put it dramatically, the infosphere is progressively absorbing any other reality. Generation Z was born onlife. Let me elaborate.

In the (fast approaching) future, more and more objects will be third-order *ITentities* able to monitor, learn, advise, and communicate with each other. A good example is provided by RFID (Radio Frequency IDentification) tags, which can store and remotely retrieve data

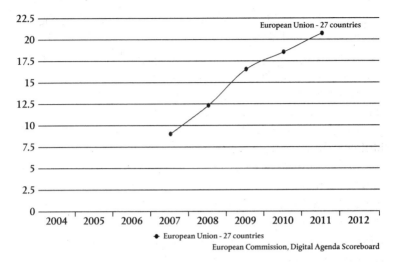

Fig. 17. In 2011, *c.*21 per cent of the EU population used a laptop to access the Internet, via wireless away from home or work.

Source: European Commission, *Digital Agenda for Europe*.

from an object and give it a unique identity, like a barcode. Tags can measure 0.4 mm^2 and are thinner than paper. Incorporate this tiny microchip in everything, including humans and animals, and you have created ITentities. This is not science fiction. According to an early report by Market Research Company In-Stat,[14] the worldwide production of RFID increased more than 25-fold between 2005 and 2010 to reach 33 billion tags. A more recent report by IDTechEx[15] indicates that in 2012 the value of the global RFID market was $7.67 billion, up from $6.51 billion in 2011. The RFID market is forecast to grow steadily over the next decade, rising fourfold in this period to $26.19 billion in 2022.

Imagine networking tens of billions of ITentities together with all the other billions of ICT devices of all kinds already available and you see that the infosphere is no longer 'there' but 'here' and it is here to stay. Nike shoes and iPod have been talking to each other since 2006, with predictable (but amazingly unforeseen) problems in terms of privacy: old models transmitted messages using wireless signals that were not encrypted and could be detected by someone else.[16] The Nest is a thermostat that learns your heating preferences. Keep using the simple dial to select a comfortable temperature, and after a week the Nest starts tuning the temperature by itself. Its sensors basically know your living patterns, habits, and preferences. And the more you interact with it, the more it learns, fine-tuning its service. Your new Samsung smart refrigerator knows what is inside it, and can offer recipe suggestions (based on the Epicurious service) as well as reminders about available fresh food and expiring items. It synchronizes with Evernote to share grocery lists. It also issues coupons. It is easy to imagine that it may learn what you like and know what you are missing, and inherit from the previous refrigerator your tastes and wishes, just as your new laptop can import your favourite settings from the old one. It could interact with your new way of cooking and with the supermarket website, just as your laptop can talk to a printer or to a smartphone. There are umbrellas that can receive wireless signals and indicate with a coloured LED whether they may be needed. Small chips in caps now help people to manage their medications with alerts, reminders, and automatic requests for refills. These are just some examples among thousands. We have all known this in theory for some time; the difference is that now it is actually happening in our kitchens.

Even money is becoming increasingly virtual. On any sterling banknote, one can still read 'I promise to pay the bearer on demand the sum of...', but the fact is that Britain abandoned the gold standard in 1931, so you should not expect to receive any precious yellow stuff in exchange. The euro, you may notice, promises absolutely nothing. Since currencies are free-floating nowadays, money may well be just a pile of digits. Indeed, when Northern Rock, a bank, collapsed, in 2007, several banks in Second Life (the online virtual world developed by Linden Lab since 2003) followed suit.[17] Players rushed to close their accounts because Second Life was not Monopoly: the exchange (technically, redemption) rate was around L$ (Linden Dollar) 260 to $1.

Likewise, in 2013, during the Cypriot bank crisis, which involved the euro this time, the virtual currency Bitcoin skyrocketed to a record high of almost $147 per Bitcoin, as people sought an allegedly safe haven for their cash.[18] That record has been broken many times since. All this is interesting because it transforms providers of in-game currencies, like Linden Lab, or indeed the Internet, into issuers of electronic money. And since the threshold between online and offline is constantly being eroded, one is left wondering when some kind of regulation will be extended to such currencies. The problem is trickier than it looks. In 2013, the FBI arrested Ross Ulbricht, allegedly the mastermind behind Silk Road, an online black market, and sought to seize his fortune of 800,000 Bitcoins, worth around $80 million.[19] For the FBI to transfer the Bitcoins out of Ulbricht's repository it needed access to the private keys protecting them (the passwords that allow one to use the Bitcoins). However, in the US scholars have argued that forcing someone to hand over their encryption keys violates the Fifth Amendment right to protection from self-incrimination.[20] When the distinction between 'money' and 'information' becomes thinner, different legal requirements may start applying.

A similar reasoning applies to fidelity cards and 'mileage' programmes. In the US, major retailers such as Best Buy and Sears have loyalty programmes that offer redeemable points, discounts, and other similar advantages. In the UK, Tesco and Sainsbury, two major retailers, run popular loyalty-card schemes. As with comparable schemes, you earn points by spending. While the money spent might not be yours (suppose your travelling expenses are reimbursed, more on this presently), the points are as good as cash. This may seem applicable only to nerds or 'desperate housewives', but even high-flyers can exchange virtual money. They just use frequent-flyer miles. According to *The Economist*, already in January 2005 'the total stock of unredeemed miles was worth more than all the dollar bills in circulation', and you can exchange them for almost anything.[21] The temptation is to pocket the miles earned through someone else's money. In 2008, for example, Britain's Parliamentary Standards watchdog complained that the Commons' Speaker, Michael Martin, had used Air Miles earned with public money for his family, the ultimate proof (Mr Martin, not the watchdog) that the UK is a hyperhistorical society.[22]

Mr Martin (born in 1945) and some of us belonging to Generation X may still consider the space of information as something we log into and log out from. Some of us may still believe that what happens online stays online. It is quite telling that Mr Martin tried to block the publication, under the Freedom of Information Act, of information about c.£5 million in yearly travel expenses by British MPs.[23] Our views about the ultimate nature of reality are still Newtonian and belong to modernity: we grew up with cars, buildings, furniture, clothes, and all sorts of gadgets and technologies that were non-interactive, irresponsive, and incapable of communicating, learning, or memorizing. However, what we still experience as the world offline is gradually becoming, in some corners of the world, a fully interactive and responsive environment of wireless, pervasive, distributed, a2a (anything to anything) information processes, that works a4a (anywhere for anytime), in real time. The day when we routinely search digitally the location of physical objects ('where are the car keys?', 'where are my glasses?'), as we already pinpoint where friends are on a map, is close. In 2008, Thomas Schmidt, Alex French, Cameron Hughes, and Angus Haines, four 12-year-old boys from Ashfold Primary School in Dorton, UK, were awarded the 'Home Invention of the Year' Prize for their Speed Searcher, a device for finding lost items. It attached tags to valuables and enabled a computer to pinpoint their location in the home.

As a consequence of the informatization of our ordinary environment, some people in hyperhistorical societies are already living onlife, in an infosphere that is becoming increasingly *synchronized*, *delocalized*, and *correlated*. Although this might be interpreted, optimistically, as the friendly face of globalization, we should not harbour illusions about how widespread and inclusive the evolution of information societies is or will be. Unless we manage to solve it, the digital divide[24] may become a chasm, generating new forms of discrimination between those who can be denizens of the infosphere, and those who cannot, between insiders and outsiders, between information-rich and information-poor. It will redesign the map of worldwide society, generating or widening generational, geographic, socio-economic, and cultural divides, between Generation Z^+ and Generation Z^-. Yet the gap will not be reducible to the distance between rich and poor countries, because it will rather cut across societies. We saw in Chapter 1 that prehistorical cultures have almost entirely disappeared, perhaps with the exception of some small tribes in remote corners of the world. The new divide will be between historical and hyperhistorical ones. We might be preparing the ground for tomorrow's informational slums.

The previous transformations already invite us to understand the world as something 'ALive'.[25] Such animation of the world will, paradoxically, make our outlook closer to that of ancient cultures, which interpreted all aspects of nature as inhabited by goal-oriented forces. We encountered a parallel phenomenon in Chapter 1, when discussing memory and the paradox of a digital 'prehistory'. The first thing Generation Z may do these days, when looking at an ICT screen, is to tap it, instead of looking for a keyboard, or wave a smartphone in front of it expecting some communication.[26] Unfortunately, such 'animation' of artefacts sometimes seems to go hand in hand with irrational beliefs about the boundless power of ICTs. When Heathrow Airport installed IRIS (Iris Recognition Immigration System), which checked registered passengers' ID by scanning their irises, one of the main problems was that some passengers tried to use the service believing that IRIS would somehow work even if they had never registered for such a service in the first place. They assumed Big Brother was already here because ICTs may easily be seen as omniscient and omnipotent gods, with minds of their own.

The next step is a rethinking of increasing aspects of reality in informational terms. It is happening before our eyes. It is becoming normal to consider the world as part of the infosphere, not so much in the dystopian sense expressed by a *Matrix*-like scenario, where the 'real reality' is still as modern and hard as the metal of the machines that inhabit it; but in the hyperhistorical, evolutionary, and hybrid sense represented by an environment such as New Port City, the fictional, post-cybernetic metropolis of *Ghost in the Shell*, 'a groundbreaking Japanese anime—the movie that gave us today's vision of cyber-space'.[27] The infosphere will not be a virtual environment supported by a genuinely 'material' world. Rather, it will be the world itself that will be increasingly understood informationally, as an expression of the infosphere. Digital third-order technologies are changing our interpretation of mechanical second- and first-order ones. At the end of this shift, the infosphere will have moved from being a way to refer to the space of information to being synonymous with reality itself.

We are changing our everyday perspective on the ultimate nature of reality from a historical and materialist one, in which physical objects and mechanical processes play a key role, to a hyperhistorical and informational one. This shift means that objects and processes are *de-physicalized*, in the sense that they tend to be seen as support-independent; consider a music file. They are *typified*, in the sense that an instance (also known as *token*) of an object—for example my copy of a music file—is as good as its type, in the example your music file of which my copy is an instance. And they are assumed to be, by default, perfectly *clonable*, in the sense that my copy and your original become indistinguishable and hence interchangeable. Given two digital objects, it is impossible to tell which one is the original source and which one is the copy just by inspecting their properties, without relying on some metadata, like a time stamp, or personal experience (you know that you made *this* copy of *that* file).

Less emphasis on the physical nature of objects and processes means that the *right of usage* is perceived to be at least as important as the *right to ownership*, with an interesting twist. It may be called *virtual materialism*. The technologies that invite 'free' usage—from social media to search engines, from online services such as free emails and messaging tools to Web 2.0 applications—rely on advertisements and hence on some data mining and on some customization of products to users. But such a reliance means that a culture of (expectations of) post-materialist, free usage of services (who would pay to have an email account or some space on a social media service?) tends to promote a market not only of services, for which you need to pay (e.g., the next holiday) but also of things that you are invited to buy (e.g., the next t-shirt). But this, in turn, favours a culture of ownership (of the t-shirt you bought). Yet, for such culture of ownership to work, potential possessions need to be commercialized as constantly renewable in order to be economically feasible (e.g., a consumer is expected to buy a new t-shirt again and again). And this closes the circle. The physical is transformed into the disposable because it is easily replaceable through free services, which are paid for by the advertisable.

When free online services promote consumerism about purchasable physical products through advertisement, the process can easily generate confusion or mistaken expectations about what is and what is not free of charge, or even whether it should be free. This confusion contributes to explaining why the more-or-less legal sharing of contents online is so popular. The Pirate Bay, the famous file-sharing site, celebrated its 10th anniversary in 2013.[28] The popularity of similar websites, which provide torrent files and links to facilitate peer-to-peer file sharing, seems more a sign of a new culture rather than damning evidence of a corrupted humanity. Anyone who argues along the materialist-historical line 'you would not steal a music CD from a shop' has not fully grasped the difficulty. Information, when treated as a commodity, has three main properties that differentiate it from other ordinary goods, including CDs and printed books. First, it is *non-rivalrous*: Alice consuming some information does not prevent Bob consuming the same information at the same time. Compare this to eating a pizza or borrowing a CD. Second, information tends to be *non-excludable*. Some information—such as intellectual properties, non-public and sensitive data, or military secrets—is often protected, but this requires a positive effort precisely because, normally, exclusion is not a natural property of information, which tends to be easily disclosed and shareable. Finally, once some information is available, the cost of its reproduction tends to be negligible (*zero marginal cost*). This is of course not the case with many goods. For all these reasons information may be sometimes seen as a *public good*, a view that in turn justifies the creation of public libraries or projects such as Wikipedia, which are freely accessible to anyone. Because of all these properties, the previous comparison with stealing a CD from a shop is not helpful. It conflates the physical with the informational. A better analogy, when it comes to downloading illegal contents, may be: 'you would not take a picture with your digital camera in an art gallery where this is not allowed'. You immediately see that this is more complicated. Indeed I would not take the picture, but grudgingly so, and if I did take a picture, I would consider this very different from stealing the corresponding postcard from the shop. Along a similar line of reasoning, from a hyperhistorical perspective, repurposing, updating, or upgrading contents need not be expressions of mere plagiarism or sloppy morality. They may be ways of appropriating and appreciating the malleable nature of informational objects.

Our society and educational system still has to catch up with such transformations. However, some new business models are already addressing such novelties by rethinking how contents are packaged and sold in the twenty-first century. In 2013, Amazon, for example, began offering buyers of printed books the corresponding ebooks for free, or at a discounted price. Called Matchbook, the scheme applies retroactively to any title bought from the store since it opened in 1995.[29] The same holds true for music files. Amazon's AutoRip service offers a free MP3 version of the CD or vinyl albums one buys.[30] It is also retroactive. Both services make the illegal exchange of digital contents much less attractive. It is a strategy that has parallels in other corners of the entertainment industry. According to Reed Hastings, Netflix's chief executive, affordable video-on-demand services may discourage people from using piracy sites because they are easier to use and of course legal and hence less risky. In an interview in 2013, he commented that 'in Canada BitTorrent is down by 50% since Netflix launched three years ago'.[31] It is an interesting comment, even if, for the sake of clarity, it must be said that BitTorrent is actually an Internet protocol,[32] like http, and as such it is widely adopted by many for perfectly legal purposes. It has become synonymous for Internet piracy only because of its widespread use in the illegal exchange of copyrighted contents.

Finally, the criterion for existence—what it means for something to be completely and ultimately real—is also changing. Oversimplifying, ancient and medieval philosophers thought that only that which is immutable, that is, God, could be said to exist fully. Anything that changes, such as an animal, moves from non-existence (there was no animal) to existence (the animal was born) back to non-existence (the animal is dead). Modern philosophers preferred to associate existence to the possibility of being subject to perception. The most empirically minded insisted on something being perceivable through the five senses in order to qualify as existing. Today, immutability and perceivability have been joined by interactability. Our philosophy seems to suggest that 'to be is to be interactable', even if that with which we interact is only transient and virtual. The following examples should help to make the previous points clearer and more concrete.

In recent years, many countries have followed the US in counting acquisition of software not as a current business expense but as an investment, to be treated as any other capital input that is repeatedly used in production over time, like a factory.[33] Spending on software now regularly contributes to GDPs. So software is acknowledged to be a (digital) good, even if somewhat intangible. It should not be too difficult to accept that virtual assets too may represent important investments.

Computing resources themselves are usually provided by hardware, which then represents the major constraint for their flexible deployment. Yet we are fast moving towards a stage when cloud computing[34] is 'softening' our hardware through 'virtualization', the process whereby one can deliver computing resources, usually built-in hardware—like a specific CPU, a storage facility, or a network infrastructure—by means of software. For example, virtualization can be adopted in order to run multiple operating systems on a single physical computing machine so that, if more machines are needed, they can be created as a piece of software—i.e., as so-called virtual machines—and not purchased as physical hardware equipment. The difference between deploying a virtual or a physical machine is dramatic. Once the virtualization infrastructure is in place, the provider of virtualized hardware resources can satisfy users' requests in a matter of minutes and, potentially, to a very large scale. Likewise, terminating or halting such a provision is equally immediate. The virtual machines are simply shut down without leaving behind any hardware component that needs to be reallocated or dismantled physically. Clearly, this will further modify our conception of what a machine is, from a historical one, based on physical and mechanical views, to a hyperhistorical one that is usage-oriented and utility-based. Dropbox, Google Documents, Apple's iCloud, or Microsoft SkyDrive have provided everyday experiences of cloud computing to millions of users for some time now. The quick disappearance of any kind of 'drive' in favour of 'ports' (USB, etc.) is a clear signal of the virtualization movement. We already met the old floppy disk drive in Chapter 1. The more recent victims are CD and DVD drives.

Next, consider the so-called 'virtual sweatshops'. These are places where workers play online games for up to twelve hours a day, to create virtual goods, such as characters, equipment, or in-game currency, or take care of the less entertaining steps in a game, for example by killing thousand of monsters to move to the next interesting level. All this and more is then sold to other players. 'Virtual sweatshops' have been with us for more than a decade. They are as old as online computer games.[35] At the time of writing, End User License Agreements (EULA, this is the contract that every user of commercial software accepts by installing it) of massively multiplayer online role-playing games (MMORPG), such as World of Warcraft, still do not allow the sale of virtual assets.[36] This would be like the EULA of MS-Office withholding from users the ownership of the digital documents created by means of the software. The situation will probably change, as more people invest hundreds and then thousands of hours building their avatars and assets. Future generations will inherit digital entities that they may want to own and be able to bequeath. Indeed, although it was forbidden, there used to be thousands of virtual assets on sale on eBay before 2007.[37] Sony, rather astutely, offers a 'Station Exchange', an official auction service that provides players with a secure method of buying and selling in dollars the right to use in game coins, items, and characters in accordance with licence agreement, rules, and guidelines. 'In its first 30 days of operation, Station Exchange saw more than $180,000 in transactions.'[38]

Once ownership of virtual assets has been legally established, the next step is to check for the emergence of property litigations. Among the oldest pieces of evidence, we find a Pennsylvania lawyer in May 2006 suing the publisher of Second Life for allegedly having unfairly confiscated tens of thousands of dollars worth of his virtual land and other property.[39]

Insurances that provide protection against risks to avatars may follow, comparable to the pet insurance you can buy at the local supermarket. Again, World of Warcraft provides an excellent example. With 11.1 million subscribers in June 2011, 10 million in October 2012, and 8.3 million in May 2013, World of Warcraft might have seen its peak.[40] Interestingly, it is being challenged by games, such as Skylanders, which are based on the onlife experience of playing with real toys that interact with the video game through a 'portal of power', that reads their tag through near-field communications technology. However, World of Warcraft is still the world's most-subscribed MMORPG. It would rank 91st in the list of 221 countries and dependent territories ordered according to population. Its users, who (will) have spent billions of man-hours constructing, enriching, and refining their digital properties, will be more than willing to spend a few dollars to insure them.

The combination of virtualization of services and virtual assets offers an unprecedented opportunity. Nowadays it is still common and easy to insure a machine, like a laptop, on which the data are stored, but not the data that it stores. This is because, although data may be invaluable and irreplaceable, it is obvious that they are also perfectly clonable at a negligible cost, contrary to physical objects, so it would be hard for an insurer to be certain about their irrecoverable loss or corruption. However, cloud computing decouples the physical *possession* of data (by the provider) from their *ownership* (by the user), and once it is the provider that physically possesses the data and is responsible for their maintenance, the user/owner of such data may justifiably expect to see them insured, for a premium of course, and to be compensated in case of damage, loss, or downtime. Users should be able to insure *their* data precisely because they own them but they do not physically possess them. 'Cyber insurance' has been around for many years,[41] it is the right thing to do, but it is only with cloud computing that it may become truly feasible. We are likely to witness a welcome shift from hardware to data in the insurance strategies used to hedge against the risk of irreversible losses or damages.

Conclusion

Despite some important exceptions—especially vases and metal tools in ancient civilizations, engravings, and then books after Gutenberg—it was the Industrial Revolution that really marked the passage from a nominalist world of unique objects to a Platonic world of types of objects. Our industrial goods are all perfectly reproducible as identical to each other, therefore indiscernible, and hence pragmatically dispensable because they may be replaced without any loss in the scope of interactions that they allow. This is so much part of our culture that we expect ideal standards and strict uniformity of types to apply even when Nature is the source. In the food industry in the UK, for example, up to 40 per cent of all the edible produce never reaches the market but is wasted because of aesthetic standards, e.g., size, shape, and absence of blemish criteria in fruit and vegetables. This because retailers know that we, the shoppers, will not buy unsightly produce.[42] Similarly, in the fashion industry, when the human body is in question, the dialectics of being uniquely like everybody else joins forces with the malleability of the digital to give rise to the common phenomenon of 'airbrushing'. Digital photographs are regularly and routinely retouched in order to adapt the appearance of portrayed people to unrealistic and misleading stereotypes, with unhealthy impact on customers' expectations, especially teenagers. The discussion of legal proposals to restrain such practices has been going on for years in France and in the UK, while evidence that warning labels and disclaimers would make a difference in the public perception is still debated.[43]

When our ancestors bought a horse, they bought *this* horse or *that* horse, not 'the' horse. Today, we find it utterly obvious and non-problematic that two cars may be virtually identical and that we are invited to test-drive and buy the model rather than an individual 'incarnation' of it. We buy the type not the token. When something is intrinsically wrong with your car, it may be a problem with the model, affecting million of customers. In 1981, the worst car recall recorded by the automobile industry so far involved 21 million Ford, Mercury, and Lincoln vehicles.[44] Quite coherently, we are quickly moving towards a commodification of objects that considers repair as synonymous with replacement, even when it comes to entire buildings.

Such a shift in favour of types of objects has led, by way of compensation, to a prioritization of informational *branding*—a process comparable to the creation of cultural accessories and personal philosophies[45]—and of *reappropriation*. The person who puts a sticker in the window of her car, which is otherwise perfectly identical to thousands of others, is fighting an anti-Platonic battle in support of a nominalist philosophy. The same holds true for the student plastering his laptop with stickers to personalize it. The information revolution has further exacerbated this process. Once our window-shopping becomes Windows-shopping and no longer means walking down the street but browsing the Web, the processes of dephysicalization and typification of individuals as unique and irreplaceable entities may start eroding our sense of personal identity as well. We may risk behaving like, and conceptualizing ourselves as, mass-produced, anonymous entities among other anonymous entities, exposed to billions of other similar individuals online. We may conceive each other as bundles of types, from gender to religion, from family role to working position, from education to social class. And since in the infosphere we, as users, are increasingly invited, if not forced, to rely on indicators rather than actual references—we cannot try all the restaurants in town, the references, so we trust online recommendations, the indicators of quality—we share and promote a *culture of proxies*. LinkedIn profiles stand for individuals, the number of linked pages stand for relevance and importance, 'likes' are a proxy for pleasant, TripAdvisor becomes a guide to leisure. Naturally, the process further fuels the advertisement industry and its new dialectics of virtual materialism. Equally naturally, the process ends up applying to us as well. In a proxy culture, we may easily be *de-individualized* and treated as a type (a type of customer, a type of driver, a type of citizen, a type of patient, a type of person who lives at that postal code, who drives that type of car, who goes to that type of restaurant, etc.). Such proxies may be further used to *reidentify* us as specific consumers for customizing purposes. I do not know whether there is anything necessarily unethical with all this, but it seems crucial that we understand how ICTs are significantly affecting us, our identities, and our self-understanding, as we shall see in Chapter 3.

3

IDENTITY

Onlife

ICTs as technologies of the self

Some time ago, I met a bright and lively graduate student, who registered with Facebook during the academic year 2003/4, when she was a student at Harvard. Her Facebook ID number was 246. Impressive. A bit like being the 246th person to land on a new planet. Such Facebook ID numbers disappeared from sight in 2009,[1] when Facebook adopted friendly usernames to make it much easier to find people. The change was necessary because, in a few years, the Facebook planet has become rather crowded, as the aforementioned student has been rapidly joined by hundreds of millions of users worldwide. Half a billion was reached in July 2010; the billion mark was passed in October 2012.

The previous story is a good reminder of how more and more people spend an increasing amount of time broadcasting themselves, digitally interacting with each other (recall the three basic operations: read/write/execute), within an infosphere that is neither entirely virtual nor only physical. It is also a good reminder of how influential ICTs are becoming in shaping our personal identities. They are the most powerful *technologies of the self*[2] to which we have ever been exposed. Clearly, we should handle them carefully, as they are significantly modifying the contexts and the practices through which we shape ourselves. Let me explain.

In the philosophy of mind, there is a well-honed distinction between who we are—let us call this our *personal identities*—and who we think we are—call this our *self-conceptions*. Needless to say, there is a crucial difference between being Napoleon and believing oneself to be Napoleon. The two selves—our *personal identities* and our *self-conceptions*—flourish only if they support each other in a mutually healthy relationship. Not only should our self-conceptions be close to, and informed by, who we really are, our actual personal identities are also sufficiently malleable to be significantly influenced by who we think we are, or would like to be. If you think you are confident, you are likely to become so, for example.

Things get more complicated because our self-conceptions, in turn, are sufficiently flexible to be shaped by who we are told to be, and how we wish to be perceived. This is a third sense in which we speak of 'the self'. It is the *social self*, so elegantly described by Marcel Proust[3] in the following passage:

> But then, even in the most insignificant details of our daily life, none of us can be said to constitute a material whole, which is identical for everyone, and need only be turned up like a page in an account-book or the record of a will; our social personality is created by the thoughts of other people. Even the simple act that we describe as 'seeing someone we know' is, to some extent, an intellectual process. We pack the physical outline of the creature we see with all the ideas we have already formed about him, and in the complete picture of him which we compose in our minds those ideas have certainly the principal place. In the end they come to fill out so completely the curve of his cheeks, to follow so exactly the line of his nose, they blend so harmoniously in the sound of his voice that these seem to be no more than a transparent envelope, so that each time we see the face or hear the voice it is our own ideas of him which we recognize and to which we listen.

The social self is the main channel through which ICTs, and especially interactive social media, exercise a deep impact on our personal identities. Change the social conditions in which you live, modify the network of relations and the flows of information you enjoy, reshape the nature and scope of the constraints and affordances that regulate your presentation of yourself to the world and indirectly to yourself, and then your social self may be radically updated, feeding back into your self-conception, which ends up shaping your personal identity. Using the previous example: if people think and say that you are confident and you wish to be seen by them as confident, then you are more likely to conceive yourself as being confident, and so you may actually become confident.

There are some classic puzzles about personal identity. They are linked to continuity through time or possible scenarios: are you the same person you were last year? Would you be the same person if you had grown up in a different place? How much of yourself would be left, if you had your brain implanted in a different body? To someone used to ruminating about such questions the whole phenomenon of the construction of personal identities online may seem frivolous and distracting, a sort of 'philosophy for dummies', unworthy of serious reflection. But in the real world, such a construction is a concrete and pressing issue to a fast-growing number of people who have lived all their adult life already immersed in Facebook, Google+, LinkedIn, Twitter, blogs, YouTube, Flickr, and so forth. To them, it seems most natural to wonder about their personal identities online, treat them as a serious work-in-progress, and to toil daily to shape and update them. It is the hyper-self-conscious generation, which facebooks, tweets, skypes, and instant-messages its subjective views and personal tastes, its private details and even intimate experiences, in a continuous flow.

Hyper-self-consciousness

Maintaining an updated and accurate presence online is not an easy task. Nor is it taken lightly. According to a study by the Pew Research Center[4] published in 2012 in the US, teenage girls send an average of 80 texts a day, followed by boys, with 'only' an average of 30. And if you thought that emails were 'so last week' because today it is all about SMS text messages, then it is time for one more upgrade. In 2012, instant messages on chat apps, such as WhatsApp, overtook SMSs for the first time, and by a wide margin: an average of 19 billion instant messages were sent daily, compared with 17.6 billion SMSs. At the time of writing, nearly 50 billion instant messages were expected to be sent per day, compared with just over 21 billion traditional SMSs.[5]

Never before in the history of humanity have so many people monitored, recorded, and reported so many details about themselves to such a large audience. The impact of so many gazillions of micro-narratives of all sorts and on all subjects is already visible. For example, they have already changed how we date and fall in love. Geosocial networking applications that allow users to locate other users within close proximity and on the basis of profiles and preferences—such as Grindr (to find, befriend, and date gay, bisexual, and bi-curious men) and Tinder (a matchmaking app that facilitates anonymous communication for dating and networking)—are popular. And according to a study conducted by the electronics retailer PIXmania in 2013,[6] tweets are the preferred way to start a relationship in the UK. It takes on average 224 tweets to start a relationship, compared to 163 text messages, 70 Facebook messages, 37 emails, or 30 phone calls. And once in a relationship, more than a third of interviewed couples admit to exchanging saucy texts and explicit pictures with each other, so-called sexting. It all starts and ends at a distance, as ICTs are also the preferred means to end a relationship: 36 per cent do it by phone, 27 per cent by text message, and 13 per cent through social media. Meeting in real life to say goodbye is so old-fashioned.

Most significantly, given the topic of this chapter, the micro-narratives we are producing and consuming are also changing our social selves and hence how we see ourselves. They represent an immense, externalized stream of consciousness, which the philosopher and psychologist William James (1842–1910) would have found intriguing:

> consciousness, then, does not appear to itself as chopped up in bits [...] it is nothing joined; it flows. A 'river' or a 'stream' are the metaphors by which it is most naturally described. In talking of it hereafter, let's call it the stream of thought, consciousness, or subjective life.[7]

Today, consciousness is still a stream (more water metaphors, recall the ones in the Preface?). But it does appear in bits, not James's bits, of course, but rather the digital ones of social media. Nothing is too small, irrelevant, or indeed private to be left untold. Any data point can contribute to the description of one's own personal identity. And every bit of information may leave a momentary trace somewhere, including the embarrassing pictures posted by a schoolmate years ago, which will disappear, of course, like everything else on this planet, but just more slowly than our former selves will.

Some Jeremiahs lament that the hyper-self-conscious Facebook generation, which is constantly asking and answering 'where are you?' on the Google map of life, has lost touch with reality. They complain that such a new generation lives in virtual bubbles where the shallowest babbles are the only currency; that it cannot engage with the genuine and the authentic; that it is mesmerized by the artificial and the synthetic; that it cannot bear anything that is slow-paced or lasts longer than a TED talk;[8] that it is made up of narcissistic, egocentric selfies (self-taken photographs usually posted online); that it is a generation incapable of responsibility because everything is expected to be erasable, revisable, and reversible anyway (one way of reading 'the right to be forgotten').

There might be some truth in all this. In 2013, Instagram contained over 23 million photos tagged #selfie, and 51 million tagged #me.[9] At the time of writing, a search engine such as Statigram indicated that the #selfie had more than doubled (52 million) and the #me almost tripled (144 million). However, in the end, I am not convinced by the Jeremiahs, for two main reasons.

First, because the supposedly genuine and the authentic, too, tend to be highly manufactured cultural artefacts. What we consider natural is often the outcome of a merely less visible human manipulation, like a well-kept garden. Indeed, we have had such an impact on our planet that geologists now speak of 'anthropocene', a topic best left for Chapter 9. 'Nature' is often how a culture understands what surrounds it.

And second, because social media also represent an unprecedented opportunity to be more in charge of our social selves, to choose more flexibly who the other people are whose thoughts and interactions create our social personality, to paraphrase Proust, and hence, indirectly, to determine our personal identities. Recall how the construction of your social self (who people think you are) feeds back into the development of your self-conception (who you think you are), which then feeds back into the moulding of your personal identity (who you are). More freedom on the social side also means more freedom to shape oneself.

The freedom to construct our personal identities online is no longer the freedom of anonymity advertised by Peter Steiner's famous cartoon, in which a dog, typing an email on a computer, confesses to another dog that 'On the Internet, nobody knows you're a dog'. Those were the nineties.[10] Today, if one is or behaves like a dog, Facebook, Google, or at least some security agency probably knows about it. Rather, it is the freedom associated with self-determination and autonomy. You may no longer lie so easily about who you are, when hundreds of millions of people are watching. But you may certainly try your best to show them who you may reasonably be, or wish to become, and that will tell a different story about you that, in the long run, will affect who you are, both online and offline. So the onlife experience is a bit like Proust's account-book, but with us as co-authors.

The Jeremiahs may still have a final point. They may be right in complaining that we are wasting a great opportunity, because, still relying on Proust's metaphor, what we are writing is not worth reading. They are disappointed by our performance as authors of our own self-narratives. But then, they have a picture of the past that is probably too rosy. Couch potatoes have been watching pictures and making small talk about their cats and the last holidays, in front of the wall of Plato's cave[11] or TV screens, well before Facebook made it embarrassingly clear that this is how most of humanity would like to spend its hard-earned free time anyway. Aristotle knew that a philosophical life requires leisure. Unfortunately, the converse is not necessarily true: leisure does not require philosophy and may easily lead only to entertainment. The result is that, as we learn from the Chorus at the beginning of *La Traviata* by Giuseppe Verdi (1813–1901):

> Giocammo da Flora.
> E giocando quell'ore volar.
> [We played at Flora's,
> And by playing, time flew.][12]

In the rest of this chapter, I shall not join the Jeremiahs. I shall not discuss how ICTs allegedly make us lonelier or are enabling us to entertain ourselves to death, until it is too late to leave Flora's party, although I will briefly come back to this point at the end of the book. I shall rather look at the slightly brighter side, and explore how the same ICTs are shaping our understanding of our selves as informational entities.

The paradox of identity

Questions about our personal identities, self-conceptions, and social selves are, of course, as old as the philosophical question 'who am I?'. So one may suspect that nothing new could sensibly be said about the topic. Yet such an attitude would be too dismissive, given the present changes. We have seen that human life is quickly becoming a matter of *onlife* experience, which reshapes constraints and offers new affordances in the development of our identities, their conscious appropriation, and our personal as well as collective self-understanding. Today, we increasingly acknowledge the importance of a common yet unprecedented phenomenon, which may be described as the online construction of personal identities. Who are we, who do we become, and who could we be, once we increasingly spend our time in the infosphere? The questions are reasonable but they hide a paradox, known as Theseus' ship. So, before addressing them, we had better have a look at the paradox itself and see whether we can avoid it.

Here is how the great ancient historian Plutarch (c. AD 46–120), describes the problem:

> [Theseus' ship] was preserved by the Athenians down even to the time of Demetrius Phalereus, for they took away the old planks as they decayed, putting in new and stronger timber in their place, insomuch that this ship became a standing example among the philosophers, for the logical question of things that grow; one side holding that the ship remained the same, and the other contending that it was not the same.[13]

You may have encountered this old problem under different disguises. Recall the axe we met in Chapter 2? Is it still your grandfather's axe, if your father replaced the handle, and you replaced the head? Theseus' ship and your grandfather's axe are systems, and it is not easy to spell out exactly what keeps them together and in working condition, as well as what makes them that particular ship and that particular axe through time and changes. The same holds true about the special system represented by you.

It seems plausible to assume that Theseus' ship, the axe, and yourself are constituted by interacting and coordinated components, but the problem concerns the changes undergone by such components. Consider your body. Most of its cells are replaced over time, yet some fundamental patterns hold, so it may not be the replacement with identical components that matters but rather that their relationship to each other and the nature of their interactions are conserved. And yet, what is this 'glue' that guarantees the unity and coordination of a system like yourself, thus allowing it to be, to persist, and to act as a single, coherent, and continuous entity in different places, at different times, and through a variety of experiences? The paradox of Theseus' ship soon starts peeping. If we wish to avoid it, we need to rely on another concept introduced in Chapter 2, that of interface.

Questions about the identity of something may become paradoxical if they are asked without specifying the relevant interface that is required to be able to answer them. Consider the following example. Whether a hospital transformed now into a school is still the same building seems an idle question to ask, if one does not specify in which context and for which purpose the question is formulated, and therefore what the right interface is through which the relevant answer may be correctly provided. If the question is asked in order to get there, for example, then the relevant interface is 'location' and the correct answer is yes, they are the same building. If the question is asked in order to understand what happens inside, then 'social function' is the relevant interface, and therefore the correct answer is obviously no, they are very different. So are they or are they not the same? The illusion that there might be a single, correct, absolute answer, independently of context, purpose, and perspective—that is, independently of the relevant interface—leads to paradoxical nonsense.

One may still retort that, even if all that I have just said is true, some interfaces should be privileged when personal identities are in question. Yet such a reply does not carry much weight. For the same analysis holds true when the entity investigated is the young Saul, who is watching the cloaks of those who laid them aside to stone Stephen,[14] or the older Paul of Tarsus, as Saul was named after his conversion. Saul and Paul are and are not the same person; the butterfly is and is not the caterpillar; Rome is and is not the same city in which Caesar was killed and that you visited last year; you are and yet you are not the same person who went there; you are and you are not your Facebook profile. It depends on why you are asking, and therefore on the right interface needed to answer the question.

This is not relativism. Given a particular goal, one interface is better than another, and questions will receive better or worse answers. The ship will be Theseus', no matter how many bits one replaces, if the question is about legal ownership. Try a Theseus trick with the taxman. However, it is already a different ship, for which the collector will not pay the same price, if all one cares about are the original planks. Questions about identity and sameness through different times or circumstances are really goal-directed questions, asked in order to attribute responsibility, plan a journey, collect taxes, attribute ownership or authorship, trust someone, authorize someone else, and so forth. Insofar as they are dealt with in absolute terms, they do not deserve to be taken seriously. For, in an allegedly purposeless and interface-free context, they make no sense, although it might be intellectual fun to play idly with them, exactly in the same way as it makes no sense to ask whether a point is at the centre of the circumference without being told what the circumference is, or being told the price of an item but not the currency in which it is given.

Our informational nature

Let us now return to our original questions: who are we, who do we become, and who could we be, as we increasingly spend our time in the infosphere? We just saw that the process of identification and reidentification of you as the same you needs to be understood in a fully informational way, through a careful analysis of the interface that is required in order to provide a reasonable answer for a specified purpose. Now, our purpose is to understand whether and how ICTs are affecting our personal identities, so the right interface seems to be offered by an informational conception of the self. And this is where the philosophy of mind can help us again. Of the many approaches that seek to characterize the nature of the self, two stand out as popular and promising for the task ahead.

One is usually dated back to the great empiricist philosopher John Locke (1632–1704). In a nutshell, your identity is grounded in the unity of your consciousness and the continuity of your memories. If this sounds a bit like Descartes it is because it follows his discussion of the 'cogito' argument: as long as you are a thinking entity, you are the specific thinking entity that is going through such specific mental processes. Allow your consciousness or memories to be hacked dramatically and you would stop being yourself. This is why you may be willing to have your mind implanted in someone else's body, but not another mind implanted in your own body.

Then there is a second approach, more recent, known as the Narrative theory of the self. According to it, your identity is a 'story', understood as a socio- and/or auto-biographical artefact. Recall what

Proust said about the social self. We 'identify' (provide identities to) each other, and this is a crucial, although not the only, variable in the complex game of the construction of personal identities, especially when the opportunities to socialize are multiplied and modified by new ICTs. Assume that everybody treats you as a completely different person every time you wake up, and you can see how you would soon go insane.

Independently of whether you prefer the Lockean or the Narrative approach, it is clear that they both provide an informational interpretation of the self. The self is seen as a complex informational system, made of consciousness activities, memories, or narratives. From such a perspective, you are your own information. And since ICTs can deeply affect such informational patterns, they are indeed powerful technologies of the self, as the following examples about embodiment, space, time, memory and interactions, perception, health, and finally education illustrate.

Embodiment: the self as an app

Informational conceptions of the self may tend to privilege a dualist view of the relationship between mind and body, more or less along the line of the distinction between hardware and software. Our culture, so imbued with ICT ideas, finds sci-fi scenarios in which you swap your old body for a new one, or the suggestion that the self may be a cross-platform structure, like an app, perfectly conceivable. Witness the debate about 'mind uploading' and 'body swap' in the philosophy of mind. It is not the funny and fictional nature of such thought experiments that is interesting here—in many cases, it tends to be distracting and fruitlessly scholastic—but the readiness with which we engage with them, because this is indicative of the particular impact that ICTs have had on how we conceptualize ourselves.

It seems indisputable that the body, its cognitive features, functions, and activities—by which I mean also our emotions, and the consciousness that accompanies them—are inextricably mixed together to give rise to a self. Our bodies and our cognition are necessary to make possible our mental lives and selves, so any form of radical dualism seems to be unjustified. Yet this truism conceals a fact and a possibility.

First, the fact. If some cause is necessary for some effect to occur, this does not mean that once the effect has indeed occurred the cause must still be there. With an analogy, there is no butterfly without the caterpillar, but it is a mistake to insist that, once the butterfly is born, the caterpillar must still be there for the butterfly to live and flourish. Likewise, our informational culture seems to look favourably to the following idea. There is no development of the self without the body, but once the latter has given rise to a consciousness, the life of the self may be entirely internal and independent of the specific body and faculties that made it possible. With another analogy, while in the air, you no longer need the springboard, even if it was the springboard that allowed you to jump so high, and your airborne time is limited by gravity.[15] All this does not mean that the self requires no physical platform. Some platform is needed to sustain the constructed self. And it does not mean that just any platform will do either. But it does open the possibility of a wider choice of platforms and of the temporary stability of a permanent self even when the platform changes.

Next, the possibility. The body itself and not just the self may also be better understood in informational terms. There are many versions of such a view, but the most popular is summarized by the 'it from bit' hypothesis formulated by the American physicist John Archibald Wheeler (1911–2008), who is probably more famous for having coined the term 'black hole'. As he put it:

> It from bit. Otherwise put, every 'it'—every particle, every field of force, even the space-time continuum itself [and therefore any body, my speci-fication]—derives its function, its meaning, its very existence entirely— even if in some contexts indirectly—from the apparatus-elicited answers to yes-or-no questions, binary choices, bits. 'It from bit' symbolizes the idea that every item of the physical world has at bottom—a very deep bottom, in most instances—an immaterial source and explanation; that which we call reality arises in the last analysis from the posing of yes–no

questions and the registering of equipment-evoked responses; in short, that all things physical are information-theoretic in origin and that this is a participatory universe.[16]

According to the 'it from bit' hypothesis, deep down our bodies too are made of information, not of some ultimate material stuff different from what is immaterial. This is not dualism but a state-based form of monism. Think of the various states in which you may find water, as vapour, liquid, or solid. If the 'it from bit' hypothesis is correct, then minds and selves on the one hand, and brains and bodies on the other, would be more like different states of information, or different informational patterns. The point that material vs. immaterial may be two states of some underlying informational stuff is reinforced by the discussion about location vs. presence.

Being in space: location vs. presence

ICTs magnify the distinction between presence and location of the self. A living organism such as a spider is cognitively present only where it is located as an embodied and embedded information-processing system. A living organism aware of its information processes, for example a dog dreaming, can be present within such processes (chasing dreamt-of rabbits) while being located elsewhere (in the house). But a self—that is, a living organism self-aware of its own information processes and its own presence within them—can choose where to be. The self, and mental life in general, is located in the brain but not present in the brain. This is why ICTs can so easily make us spend so much of our conscious time present elsewhere from where we are bodily located.

Being in time: outdating vs. ageing

ICTs increase the endurance effect, for in digital environments it is easier to identify and reidentify exactly the same thing through time. The problem is that the virtual may or may not work properly, it may be old or updated, but it does not grow old; it 'outdates', it does not age. If you think of it, nothing that outdates can outdate more or less well. By contrast, the self ages, and can do so more or less graciously. The effect, which we have only started to experience and with which we are still learning to cope, is a chronological misalignment between the self and its online habitat, between parts of the self that age (e.g. my face) and parts that simply outdate (e.g. the picture of my face on my driving licence). Asynchronicity is acquiring a new meaning in onlife contexts.

Memories and interactions: pegging the self

We saw that memory plays a crucial role in the construction of personal identity. Obviously, any technology, the primary goal of which is to manage memories, is going to have an immense influence on how individuals develop and shape their own personal identities. It is not just a matter of mere quantity. The quality, availability, accessibility, and replaying of personal memories may deeply affect who we think we are and may become. The Korean War was, for example, the first major conflict with a soundtrack: soldiers could be listening to the same songs at home, in the barracks, or during a battle.[17] Similar 'repeatable' memories cannot but have a deep impact on how individuals exposed to them shape their understanding of their past, the interpretation of what has happened to them, and hence how they make sense of who they are. Generation X was the first ubiquitous 'replay' generation. Today, our madeleines are digital.[18]

Until recently, the optimistic view was that ICTs empowered individuals to mould their personal identities. The future seems more nuanced. Recorded memories tend to freeze and reinforce the nature of their subject. The more memories we accumulate and externalize, the more narrative constraints we provide for the construction and development of our personal identities. Increasing our memories also means decreasing the degree of freedom we might enjoy in redefining ourselves. Forgetting is part of the process of self-construction.

A potential solution, for generations to come, may be to be thriftier with anything that tends to crystallize the nature of the self, and more adept in handling new or refined skills of self-construction. Capturing, editing, saving, conserving, managing one's own memories for personal and public consumption will become increasingly important not just in terms of protection of informational privacy, as we shall see in Chapter 4, but also in terms of construction of a healthier personal identity. The same holds true for interactions. The onlife experience does not respect boundaries between different online and offline environments, with the result that, as I have already mentioned, the scope for naïve lying about oneself on Facebook is increasingly reduced. In this case, the solution may lie in the creation of more affordances and safer spaces for self-expression and self-construction (see for example Diaspora, the open source Facebook).

Perception: the digital gaze

The gaze is a composite phenomenon, with a long and valuable tradition of analyses.[19] The idea is rather straightforward: it is comparable to seeing oneself as seen by others, by using a mirror ('what do people see when they see me?'). Note, however, that it should not be confused with seeing oneself in a mirror (ego surfing or vanity googling). Rather, the self observes 'the observation of itself' by other selves (including, or sometimes primarily itself) through some medium. In child development, the gazing phase is theorized as a perfectly healthy and normal stage, during which the individual learns to see her- or himself by impersonating, for example, a chair ('how does the chair see me?'), or simply placing her- or himself in someone else's shoes.

The digital gaze is the transfer of such phenomenon to the infosphere. The self tries to see itself as others see it, by relying on ICTs that greatly facilitate the gazing experience. In the end, the self uses the digital representation of itself by others in order to construct a virtual identity through which it seeks to grasp its own personal identity (the question 'who am I for you?' becomes 'who am I online?'), in a potentially recursive feedback loop of adjustments and modifications leading to an onlife equilibrium between the offline and the online selves. The observing process is normally hidden and certainly not advertised. And yet, by its very nature, the digital gaze should be understood both as an instance of presumed common knowledge of the observation—we already met this concept; it is what happens when I know that you know that I know etc.... that this is the way I am seen by you—and as a private experience—it is still *my* seeing of myself, even if I try to make sure that such seeing is as much like your seeing as I can. The digital translation of the gaze has important consequences for the development of personal identities.

First, there is the amplification, postponement (in terms of age), and prolongation (in terms of duration) of the gazing experience. This means that the tendency of the gaze to modify the nature of the self that is subject to it becomes a permanent feature of the onlife experience. The hyperconscious self never really stops trying to understand how it is seen by the other. Second, through the digital gaze, the self sees itself from a third-person perspective through the observation of itself in a proxy constrained by the nature of the digital medium, one that provides only a partial and specific reflection. It would be like being constrained to look at oneself from a distorting mirror that can provide no access to other images of oneself. Third, the more powerful, pervasive, and available ICTs are, the more the digital gaze may become mesmerizing: one may be lost in one's own perception of oneself as attributed by others in the infosphere. And finally, the experience of the digital gaze may start from a healthy and wilful exposure/exploration by the self of itself through a medium, but social pressure may force it on selves that are then negatively affected by it, leading them to modify themselves in a way that could impose some external and alien rules on the process of construction of one's own identity. If you see me seeing you in a way that you do not like, you may be tempted to adapt and modify your self until the way you see me seeing you finally pleases you, and this may not be necessarily healthy.

Bodies of information: e-health

Monsieur Homais is one of the less likeable characters in *Madame Bovary*, the famous novel by Gustave Flaubert (1821–80). The deceitful pharmacist fakes a deep friendship for Charles Bovary, Emma's husband. In fact, he constantly undermines his reputation with his patients, thus contributing to Charles's ruin. Monsieur Homais is not merely wicked. A smart man, he has been convicted in the past for practising medicine without a licence. So he worries, reasonably, that Charles might denounce him to the authorities for the illicit business of health advice and personal consultations that he keeps organizing in his pharmacy. The ultimate success of the pharmacist's dodgy schemes is not surprising. Those were the days when blacksmiths and barbers could regularly act as dentists and surgeons (after all, Charles is not a doctor either, but only a 'health officer'). Patients and doctors had to meet face to face in order to interact. And access to health information was the privilege of the few. Mail and telegraph messages were, of course, commonly available, but neither allowed real-time conversations.

Madame Bovary was serialized in 1856, exactly twenty years before Alexander Graham Bell (1847–1922) was awarded a patent for the electric telephone by the United States Patent and Trademark Office. Once ICTs of all kinds began to make possible quick consultations and rapid responses, being 'on call' acquired a new meaning, telemedicine was born, and the Messieurs Homais around the world started to find it increasingly difficult to make a living. Behind the success of ICT-based medicine and well-being lie two phenomena and three trends, which are consistent with the dephysicalization and typification of individuals that we have encountered in Chapter 2.

The first phenomenon. This may be labelled 'the transparent body'. ICTs enable us to measure, model, simulate, monitor, and manage our bodies ever more deeply, accurately, and non-invasively. So they are essential to prevent or treat an ever-increasing variety of diseases.

Unsurprisingly, global revenue for consumer medical devices is expected to grow steadily in the years ahead, as shown in Figure 18.

The phenomenon of the 'transparent body' is related not only to illnesses but also to well-being in general. According to a report by Juniper Research,[20] the global market of wearable smart devices, which monitor our sport activities and levels of fitness and can suggest new training programmes, will grow from $1.4 billion of sales in 2013 to $19 billion by 2018. The battle for your wrist between Adidas' Micoach, Nike's Fuelband, Motorola's Motoactv, and other similar gadgets has a powerful health-related underpinning. It may be interpreted as a competition for who will succeed in making our bodies more usefully and pleasantly transparent to ourselves. ICTs are making us more easily explorable, have increased the scope of possible

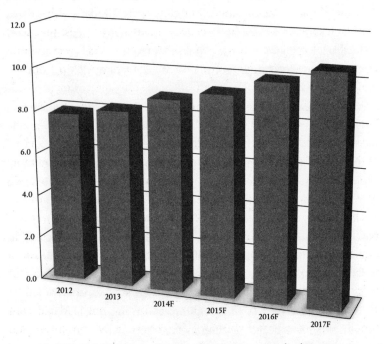

Fig. 18. Global Revenue Forecast for Consumer Medical Devices.
Data source: IHS Inc., September 2013.

interactions from without and from within our bodies (nanotechnology), and made the boundaries between body and environment increasingly porous, from X-ray to fMRI. From being black boxes, we are quickly becoming white boxes through which we can see.

The second phenomenon. This may be labelled 'the shared body'. 'My' body can now be easily seen as a 'type' of body, thus easing the shift from 'my health conditions' to 'health conditions I share with others'. And it is more and more natural to consider oneself not only the source of information (what you tell the doctor) or the owner of information about oneself (your health profile), but also a channel that transfers DNA information and corresponding biological features between past and future generations (you are the biological bridge between your parents and your children). Rapid genetic testing is now easily available for $99.

Some of the obvious advantages of 'the shared body' are that there may be less loneliness, more hope, easier spreading of best practices, more prevention, and better planning. A serious risk is the 'everybody does it' factor: we may find normality in numbers, and shift from the medicalization to the socialization of unhealthy choices or habits. If I can join a group that endorses nail-biting, I may end up thinking it is not an impulse control disorder that needs treatment. Interestingly, these phenomena feed into the hyperconscious issue ('whose identity?') we met earlier, the information privacy problem ('whose information?') that I shall discuss in Chapter 5, as well as the empowering possibilities ('whose options?') I shall analyse in Chapters 6 and 7.

The 'transparent body' and the 'shared body' are correlated to three main trends: a democratization of health information, an increasing availability of user-generated contents that are health-related, and a socialization of health conditions. Democratization means here that more and more information is available, accessible to, and owned by an increasing number of people. But patients are not just avid consumers of medical information, they are also active producers and sharers of large quantities of health-related contents. The 'wikification' of medical information is already a significant phenomenon with global consequences.[21] As a further consequence, we are witnessing an unprecedented socialization of health conditions. I have already mentioned a major risk. The advantages can be equally important. You only need to check 'multiple sclerosis' on YouTube, for example, to appreciate how easily and significantly ICTs can shape and transform our sense of belonging to a community of patients and carers.

Given the previous analysis, it is obvious why, already in 2001, the Kennedy Report in the UK stated that

> All health care is information driven, so the threat associated with poor information is a direct risk to the quality of healthcare service and governance in the NHS [National Health Service].

By 2018, the world population will consist of more people over 65 than children under 5, for the first time in the history of humanity (see Figure 19).[22]

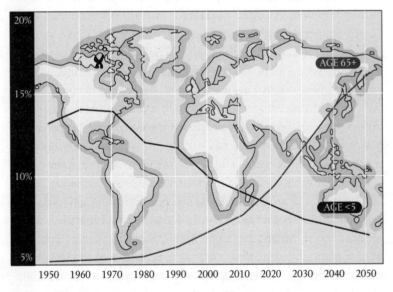

Fig. 19. An Ageing World.
Source: Unknown

We are getting older, more educated, and wealthier, so e-health can only become an increasingly common, daily experience, one of the pillars of future medical care, and obviously a multi-billion-dollar business, some of which will inevitably be dodgy. Your inbox is full of dubious medical advices and pharmaceutical products. Which of course leads us back to Monsieur Homais. Emma learns from him how to acquire the arsenic with which she will commit suicide. During her horrible agony, her husband desperately 'tried to look up his medical dictionary, but could not read it'. Nowadays you only need the usual Wikipedia. Just check under 'Arsenic poisoning'. You will find there both diagnosis and treatment.

e-ducation

The last topic I wish to touch upon in this chapter deserves more space. Few things influence us as much as our interactions with each other. This is even more so when such interactions are pedagogical. The idea is not new. What is new is the challenge we are facing when pedagogical interactions occur in hyperhistorical societies, onlife.

Perhaps there were times when 'civilized', 'cultured', and 'educated' could rightly be treated as synonymous. Thucydides (c.460–c.395 BC) and Cicero (106–43 BC) may come to mind. Some characters in Jane Austen (1775–1817), Henry James (1843–1916; brother of William James, whom we met earlier), or Edith Wharton (1862–1937) seem to draw little distinction between the three corresponding concepts. Yet today they hardly overlap at all. 'Civilized' refers to a person's manners and behaviours. 'Cultured' qualifies someone who is engaged with arts, letters, and other intellectual pursuits. And 'educated' is usually applied to people who have successfully attended learning or training courses offered by primary, secondary, or tertiary (higher) institutions. One could be any of the three without being either of the remaining two.

Globalization has greatly contributed to this differentiation, even if it has been pushing it in opposite directions, local and global. Michel de Montaigne (1533–92) already knew that 'civilized' and 'cultured' had local interpretations. The difference is that today we feel increasingly less justified in prioritizing one 'locality' over the others, be this Rio de Janeiro, New Delhi, Beijing, or Tokyo. We know that it is a matter of civilized manners either to take off one's shoes or to keep them on, depending on where we are or whom we are visiting. We accept that Alice may be cultured even though she has no clue about bossa nova music, Sattriya dance, Sichuanese opera, or Noh theatre. Yet education is not necessarily about any of this. Compulsory schooling, the institutionalization of teaching and learning, universal pedagogical principles, and the globalization of the job market began detaching education from upbringing a long time ago. Today, an avionics engineer, a scholar of Mexican literature, a developmental psychologist, or a macroeconomist, are increasingly evaluated on global, international standards.

ICTs have further magnified and accelerated both trends. The more they expose us to each other, the more they make us aware that being civilized or cultured is a *relative* matter. The infosphere has many nodes but no ultimate centre, so one can only be more or less provincial. But by making us share needs and expectations on a global scale, ICTs also make us demand similar sets of minimal educational standards. In the infosphere, being educated is increasingly a de-localized, uniform, and global phenomenon. It is not a relative but a *relational* matter, in the following sense.

Education is largely about transmission of knowledge and of how it may be increased. Broadly construed, the knowledge in question includes not only the critical acquisition of facts and formulae, but also understanding, and the appreciation of values and interpretations, ways of living and traditions, abilities and skills. The previous list is incomplete but, in each case, education inevitably relates the educated to something else. The transmitted knowledge may be *of* a place or a practice, *that* such-and-such is or is not the case, of *why* it may or may not be otherwise, or *about how* something works. The solidity of the relation between the educated and the knowledge that is being transmitted is ultimately tested by truth-tracking and truth-generating features. No matter when or where Alice lived, she cannot be said to know Los Angeles, if she has never been there and thinks it is a small Italian village; she cannot be said to know that the Earth is flat; and she cannot be said to know how to drive a car, if she has never driven one in her life.

Of course, history or geography taught in China may be quite different from history or geography taught in Japan or South Korea,[23] and some biology courses in the United States may not always be comparable to the same courses taught in Europe.[24] So Alice and Bob may be educated quite differently, relatively to the contexts in which they were brought up. However, this is not the point. The point is that expectations about being cultured and civilized ought to be carefully relativized, whereas expectations about being educated ought to be absolute. This is why it makes sense to compare the success of school-children in different countries, but not their level of cultural sophistication, and the quality of universities in the world, but not the degree of the civilized refinement of their students.

Since education is coupled to knowledge, when the latter changes, the former must follow suit. Now, the information society has witnessed the fastest growth of knowledge in the history of humanity. It is a growth that is qualitative and quantitative, both in scope and in pace. In Chapter 1, we saw some evidence in terms of the immense growth of available data. Unsurprisingly, the exponential increase of what may be transmitted has caused a major crisis in how we conceive education and organize our pedagogical systems. One widespread and popular reaction has been to try to transform ICTs from being part of the problem to being part of the solution. This is valuable but also distracting. The real educational challenge in hyperhistorical societies is increasingly *what* to put in the curriculum not *how* to teach it.

The *how* is easy, not because it is straightforwardly feasible, but because it is more clearly understood. Digital technologies in the classroom are an old phenomenon. A century after Turing's birth, universities are rushing to put their courses online, and the market of e-learning is blooming. There is much to be said in favour of (distance) e-learning, when it is not a form of 'unmanned teaching' or merely cheap outsourcing. As its supporters rightly stress, it has made a vast reservoir of educational contents available to millions of people, and it promises to deliver even more to ever more. ICTs may allow a degree of didactic customization unprecedented in non-elitist contexts: the personalization of the educational experience for millions of individuals. But all this is a matter of delivery policies, methods, techniques, and technologies. If it is taken to be a solution of how to educate Generation Z and the others which will follow, then we are mistaking a painkiller for a cure. The real headache is not the *how*. Since the late eighties we have become enthusiastic about MOOs (text-based online virtual reality systems for multiple users connected at the same time), literary hypertexts, glove-and-goggle VR (virtual realities), HyperCard, Second Life, and now MOOCs (massive open online courses). More fashions and further acronyms will certainly follow. Yet the real headache is the *what*.

There is no clear and fixed answer to the educational *what*-question in hyperhistorical societies. Not only because we have never been here before, but also because, as in the past, the answer still depends on the answer to another question: what education is for. Nevertheless, a few considerations may delimit the space within which we can search for a solution. To introduce them, let me use a simple example.

Suppose Alice is playing a computer game. There are things that she knows, such as that there is a monster hiding. This is her knowledge. There are other things that she knows that she does not know, such as where the monster is hiding, and that is why she is searching for it. This is her lack of knowledge or simply insipience. Then there are further things that she is not quite sure she knows, such as whether her weapons are sufficiently powerful to kill the monster, and that is why she is trying to acquire some more. This is her uncertainty. And finally, there are things that she does not even know that she does not know: there is a magic sword that can kill the monster. This is her ignorance. We can translate the example into informational terms, thus:

1. knowledge: information Alice has (there is a monster)
2. insipience: information that Alice is aware she is missing (where is the monster hiding?)
3. uncertainty: information about which Alice is uncertain (are my weapons sufficient to kill the monster?)
4. ignorance: information that Alice is not aware she is missing (if only she knew that she is missing the fact that there is a magic weapon!)

Education has always had the goal of increasing (1) and decreasing (2), (3), and (4).

Regarding (1), in a world awash with easily accessible information, cheap ICTs, and a plentiful intellectual workforce, increasing basic knowledge has become easy, hence the success of MOOCs based on interactive participation and open access through the Web. The educational problem with (1) is that new information always requires some old background information to become meaningful and useful, and to be appropriated critically. So we need to understand how much and what kind of background information—things one needs to know, independently of whether one may check them on Wikipedia if necessary—Alice needs to acquire in order to be educated today.

Regarding (2), education should teach us the limits of our knowledge, what kind of information we do not have but might want to acquire, and hence a good taste for the right sort of questions we should ask. We are all insipient: it is how we handle our degree of insipience that makes a difference. So, the educational problem with (2) becomes which kind of unknowns Alice should be taught to be aware of today.

Regarding (3), education should teach us to be careful about what we think we know, and hence the art of doubting and being critical even of the seemingly certain. We are all fallible, it is how we handle our degree of fallibility that makes a difference. So, the educational problem with (3) becomes what kind of uncertainties Alice should be taught today.

As for (4), it is an internal problem. This is why only we can describe it for Alice. If Alice knew what she does not know that she is missing, she would be insipient or uncertain about it, not ignorant, after all. Now, imagine that we could talk to Alice: at a stroke, we could tell her that she is missing some information about the existence of a magic sword, so that particular instance of her ignorance would be erased. This is what a more globalized e-ducation across geographical borders and academic boundaries can do. It cannot erase humanity's ignorance, but it can place each human being on one side of the same divide, even if, by definition, we, as humanity, do not know where that divide is. Let me explain using the same example.

Suppose Bob knows that he does not know where the magic sword is, but he is not even aware of missing the information that there is a monster nearby. If Alice and Bob share their insipience, then they can decrease their ignorance: together they will know that they do not know both where the monster is and where the magic sword is. It may sound funny, but this is a great improvement. Internal ignorance is decreased, even if external ignorance may not be (what Alice and Bob as a group are unaware that they are missing—imagine both of them being ignorant about the existence of a friendly wizard).

In what I have said so far, the tension between facts and skills remains. Is it more important to teach Alice that the monster has seven heads, all of which need to be cut in order to kill it, or to teach Alice how to cut them? You immediately see the misleading nature of the facts vs. skills dichotomy. She needs to have both kinds of knowledge, or she will not win the game. However, today, because so much information is just a click away, there seems to be a tendency to privilege know-how over know-that. This is silly, especially if one recalls the importance of background information stressed earlier. It is also misleading, if by privileging know-how we promote a culture of only users and consumers, instead of a culture of designers and producers as well. The information society is a neo-manufacturing society in which information is both the raw material we produce and manipulate and the finished good we consume. In such a society, when it comes to skills, we really need to place more emphasis on the so-called 'maker's knowledge', the knowledge that is enjoyed by those who know how to design and produce the artefacts, that is, those who know how to create, design, and transform information. This is easier said than done, because our Western culture is based on a deeply ingrained Greek divide between *episteme* (science and 'knowledge that'), which is highly valued and respected, and *techne* (technology and 'knowledge how'), which is seen as secondary. Think of how 'vocational' skills and training are evaluated in our society. As we just saw with Alice and the monster, this is a false dichotomy. It is also one that focuses too much on the wrong side of the coin. Using our previous example, the game of knowledge includes players, watchers, and designers. A fact-based education and a skill-based education are strategies for players. They both address Alice as a user, not as a producer of information. The risk is to develop a 'luxury box' reaction, with watchers enjoying the knowledge game without actually playing. It used to be called an ivory tower. Meanwhile, an important part of the real business of education takes place at the game-designer level.

We need to teach Alice-the-user how to play the game of knowledge successfully, Alice-the-intellectual how to observe and study the game critically, and Alice-the-designer how to devise the whole game properly. So the question becomes: what sort of abilities should we privilege and teach to tomorrow's curators, producers, and designers of information? The answer seems to me quite obvious: the languages through which information is created, manipulated, accessed, and consumed. By this I do not mean only one's own mother tongue, the full mastery of which is the first, basic, necessary step for any other form of education. I also mean English (or whatever language will one day be the international medium of communication), mathematics, programming, music, graphics, and all those natural and artificial languages in which Alice and the new generations need to be proficient at an early stage of their development, in order to be able to understand critically the accessible information, create and design new information, and share it with others.

ICTs are great in making information available; they are less successful in making it accessible, and even less so in making it usable. Try reading a scientific entry (pick your science) on Wikipedia and the likelihood is that much of it will be impenetrable, if you do not speak the right language. More availability and better accessibility are issues on the side of the providers. But information production and design at the beginning of the availability process, and usability and comprehension of the accessible information at the end, are issues that involve Alice's education. Languages are best learnt when we are young. And language proficiency is not a matter of memorized facts or practised skills, but of finely tuned abilities. Alice needs to learn the languages of information as early as possible.

Conclusion

In this chapter and in Chapters 1 and 2, I have sketched how ICTs have brought about some significant transformations in our history (hyperhistory), in our environment (infosphere), and in the development of our selves (the onlife experience). At the roots of such transformations, there seems to be a deep philosophical change in our views about our 'special' place and role in the universe. It is a fourth revolution in our self-understanding, as I shall argue in Chapter 4.

4

SELF-UNDERSTANDING

The Four Revolutions

The first three revolutions

S cience changes our understanding in two fundamental ways. One
may be called *extrovert*, or about the world, and the other *introvert*,
or about ourselves. Three scientific revolutions in the past had great
impact both extrovertly and introvertly. In changing our understanding of the external world, they also modified our conception of who
we are, that is, our self-understanding. The story is well known, so
I shall recount it rather quickly.

We used to think that we were at the centre of the universe, nicely
placed there by a creator God. It was a most comfortable and reassuring
position to hold. In 1543, Nicolaus Copernicus (1473–1543) published his
treatise on the movements of planets around the sun. It was entitled
On the Revolutions of Celestial Bodies (*De Revolutionibus Orbium Coelestium*).
Copernicus probably did not mean to start a 'revolution' in our self-
understanding as well. Nonetheless, his heliocentric cosmology for ever
displaced the Earth from the centre of the universe and made us
reconsider, quite literally, our own place and role in it. It caused such
a profound change in our views of the universe that the word 'revolution' begun to be associated with radical scientific transformation.

We have been dealing with the consequences of the Copernican
revolution since its occurrence. Indeed, it is often remarked that one of

the significant achievements of our space explorations has been an external and comprehensive reflection on our human condition. Such explorations have enabled us to see Earth and its inhabitants as a small and fragile planet, from outside. Of course, this was possible only thanks to ICTs. Figure 20 reproduces what is probably the first picture of our planet, taken by the US satellite Explorer VI on 14 August 1959.

After the Copernican revolution, we retrenched by holding on to the belief in our centrality at least on planet Earth. The second revolution occurred in 1859, when Charles Darwin (1809–82) published his *On the Origin of Species by Means of Natural Selection, or the*

Fig. 20. First Picture of Earth, taken by the US Satellite Explorer VI. It shows a sunlit area of the central Pacific Ocean and its cloud cover. The signals were sent to the South Point, Hawaii tracking station, when the satellite was crossing Mexico.

Courtesy of NASA, image number 59-EX-16A-VI, date 14 August 1959.

Preservation of Favoured Races in the Struggle for Life. In his work, Darwin showed that all species of life have evolved over the years from common ancestors through natural selection. This time, it was the word 'evolution' that acquired a new meaning.

The new scientific findings displaced us from the centre of the biological kingdom. As with the Copernican revolution, many people find it unpleasant. Indeed, some people still resist the idea of evolution, especially on religious grounds. But most of us have moved on, and consoled ourselves with a different kind of importance and a renewed central role in a different space, one concerning our mental life.

We thought that, although we were no longer at the centre of the universe or of the animal kingdom, we were still the masters of our own mental contents, the species completely in charge of its own thoughts. This defence of our centrality in the space of consciousness came to be dated, retroactively and simplistically, to the work of René Descartes (1596–1650). His famous 'I think therefore I am' could be interpreted as also meaning that our special place in the universe had to be identified not astronomically or biologically but mentally, with our ability of conscious self-reflection, fully transparent to, and in control of, itself. Despite Copernicus and Darwin, we could still regroup behind a Cartesian trench. There, we could boast that we had clear and complete access to our mental contents, from ideas to motivations, from emotions to beliefs. Psychologists thought that introspection was a sort of internal voyage of discovery of mental spaces. William James still considered introspection a reliable, scientific methodology. The mind was like a box: all you needed to do to know its contents was to look inside.

It was Sigmund Freud (1856–1939) who shattered this illusion through his psychoanalytic work. It was a third revolution. He argued that the mind is also unconscious and subject to defence mechanisms such as repression. Nowadays, we acknowledge that much of what we do is unconscious, and the conscious mind frequently constructs reasoned narratives to justify our actions afterwards. We know that we cannot check the contents of our minds in the same way we search the contents of our hard disks. We have been displaced from the centre of the realm of pure and transparent consciousness. We acknowledge being opaque to ourselves.

There are, of course, serious doubts about psychoanalysis as a scientific enterprise. Yet, one may still be willing to concede that, culturally, Freud was influential in initiating the radical displacement from our Cartesian certainties. What we mean by 'consciousness' has never been the same after Freud, but we may owe him more philosophically than scientifically. So you may prefer to replace psychoanalysis with contemporary neuroscience as a more likely candidate for such a revolutionary scientific role. Either way, today we acknowledge that we are not immobile, at the centre of the universe (Copernican revolution), that we are not unnaturally separate and diverse from the rest of the animal kingdom (Darwinian revolution), and that we are far from being Cartesian minds entirely transparent to ourselves (Freudian or neuroscientific revolution).

One may easily question the value of the interpretation of these three revolutions in our self-understanding. After all, Freud himself was the first to read them as part of a single process of gradual reassessment of human nature. His interpretation was, admittedly, rather self-serving. Yet the line of reasoning does strike a plausible note, and it can be rather helpful to understand the information revolution in a similar vein. When nowadays we perceive that something very significant and profound is happening to human life, I would argue that our intuition is once again perceptive, because we are experiencing what may be described as a fourth revolution, in the process of dislocation and reassessment of our fundamental nature and role in the universe.

The fourth revolution

After the three revolutions, was there any space left where we could entrench ourselves smugly? The French philosopher and theologian

Blaise Pascal (1623–62) had poetically suggested one. In a famous quote, he had remarked that

> Man is but a reed, the feeblest thing in nature, but he is a thinking reed. The entire universe need not arm itself to crush him. A vapour, a drop of water suffices to kill him. But, if the universe were to crush him, man would still be nobler than that which killed him, because he knows that he dies and the advantage that the universe has over him, the universe knows nothing of this. *All our dignity then, consists in thought. By it we must elevate ourselves, and not by space and time which we cannot fill.* [emphasis added][1]

Centuries later, Pascal's dignity of thought remained still unchallenged by the three revolutions we encountered earlier. We could still hold on to the view that our special place in the universe was not a matter of astronomy, biology, or mental transparency, but of superior thinking abilities. This was the implicit line of defence of our exceptional place in the universe which was still standing. Intelligence was, and still is, a rather vague property, difficult to define, but we were confident that no other creature on Earth could outsmart us. Whenever a task required some intelligent thinking, we were the best by far, and could only compete with each other. We thought that animals were stupid, that we were smart, and this seemed the reassuring end of the story. We quietly presumed to be at the centre of the infosphere, joined by no other earthly creature.

It was a dangerous line of defence, which, ironically, Pascal himself helped to undermine. In 1645, he published a short 'Dedicatory Letter' to Pierre Séguier (1588–1672), the chancellor of France. The name may ring a bell because Séguier appears in *The Three Musketeers*, handling a delicate but different letter. The document, entitled *Arithmetical Machine (Machine d'arithmétique)*,[2] described a new computational device, which Pascal had built for his father to help him deal with the tiresome calculations required by his job as a supervisor of taxes at Rouen. Thanks to some clever solutions,[3] the machine could perform the four arithmetical operations quite well. Known today as 'Pascalina', it became the only functional mechanical calculator in the 17th century.

It was a success, and nine still survive. It had an enormous influence on the history of calculators and on Gottfried Leibniz (1646–1716), the great German mathematician and philosopher who devised the modern binary number system and is rightly considered the first computer scientist and information theorist. In the letter, Pascal wrote:

> Dear reader, this notice serves to let you know that I present to the public a small machine of my invention, by means of which you can, without any trouble, do all the operations of arithmetic, and relieve you of the work that has often tired your mind (*esprit*), when you operate by using a token (*jeton*)[4] or the pen.[5]

Maybe because he was a religious man, Pascal did not see any inconsistency between his view that 'All our dignity then, consists in thought' and the arithmetical abilities of his machine. He could envision only a fruitful collaboration between his father and his Pascalina. It was left to another philosopher, on the other side of the channel, to provide the missing link.

Six years after the publication of Pascal's letter, in 1651, Thomas Hobbes (1588–1679), one of the most influential political thinkers of all time, published his masterpiece: *Leviathan or The Matter, Forme and Power of a Common Wealth Ecclesiasticall and Civil*.[6] Not exactly a book where you would expect to find the roots of our information society, and yet, in chapter 5, a groundbreaking idea entered into our culture for the first time:

> For 'reason' in this sense is nothing but 'reckoning,' that is adding and subtracting, of the consequences of general names agreed upon for the 'marking' and 'signifying' of our thoughts; I say 'marking' them when we reckon by ourselves, and 'signifying' when we demonstrate or approve our reckonings to other men.

Thinking was reasoning, reasoning was reckoning, and reckoning could already be done by a Pascalina. The seeds of the fourth revolution had been sown. Future generations of Pascalina were going to relieve us not just of our mentally tiring work, but also of our central role as the only smart agents in the infosphere.

Pascal had not considered the possibility that we would engineer autonomous machines that could surpass us at processing information logically and hence be behaviourally smarter than us whenever information processing was all that was required to accomplish a task. The oversight became clear with Alan Turing's work, the father of the fourth revolution.

Turing displaced us from our privileged and unique position in the realm of logical reasoning, information processing, and smart behaviour. We are no longer the undisputed masters of the infosphere. Our digital devices carry out more and more tasks that would require some thinking from us if we were in charge. We have been forced to abandon once again a position that we thought was 'unique'. The history of the word 'computer' is indicative. Between the seventeenth and the nineteenth century, it was synonymous with 'a person who performs calculations' simply because there was nothing else in the universe that could compute autonomously. In 1890, for example, a competitive examination for the position of 'computer' by the US Civil Service had sections on 'orthography, penmanship, copying, letter-writing, algebra, geometry, logarithms, and trigonometry'.[7] It was still Hobbes's idea of thinking as reckoning. Yet by the time Turing published his classic paper entitled 'Computing machinery and intelligence',[8] he had to specify that, in some cases, he was talking about a '*human* computer', because by 1950 he knew that 'computer' no longer referred only to a person who computes. After him, 'computer' entirely lost its anthropological meaning and of course became synonymous with a general-purpose, programmable machine, what we now call a Turing machine.

After Turing's groundbreaking work, computer science and the related ICTs have exercised both an extrovert and an introvert influence on our understanding. They have provided unprecedented scientific insights into natural and artificial realities, as well as engineering powers over them. And they have cast new light on who we are, how we are related to the world and to each other, and hence how we conceive ourselves. Like the previous three revolutions, the fourth revolution removed a misconception about our uniqueness and also provided the conceptual means to revise our self-understanding. We are slowly accepting the post-Turing idea that we are not Newtonian, stand-alone, and unique agents, some Robinson Crusoe on an island. Rather, we are informational organisms (*inforgs*), mutually connected and embedded in an informational environment (the infosphere), which we share with other informational agents, both natural and artificial, that also process information logically and autonomously. We shall see in Chapter 6 that such agents are not intelligent like us, but they can easily outsmart us, and do so in a growing number of tasks.

Inforgs

I mentioned earlier that we are probably the last generation to experience a clear difference between online and offline environments. Some people already spend most of their time onlife. Some societies are already hyperhistorical. If home is where your data are, you probably already live on Google Earth and in the cloud. Artificial and hybrid multi-agents, i.e., partly artificial and partly human (consider, for example, a bank), already interact as digital agents with digital environments and, since they share the same nature, they can operate within them with much more freedom and control. We are increasingly delegating or outsourcing to artificial agents our memories, decisions, routine tasks, and other activities in ways that will be progressively integrated with us. All this is well known and relevant to understanding the displacement caused by the fourth revolution, in terms of what we are not. However, it is not what I am referring to when talking about *inforgs*, that is, what the fourth revolution invites us to think we may be. Indeed, there are at least three more potential misunderstandings against which I should warn you.

First, the fourth revolution concerns, negatively, our newly lost 'uniqueness' (we are no longer at the centre of the infosphere) and, positively, our new way of understanding ourselves as inforgs. The fourth revolution should not be confused with the vision of a 'cyborged' humanity. This is science fiction. Walking around with something like a Bluetooth wireless headset implanted in your ear does not seem the best way forward, not least because it contradicts the social message it is also meant to be sending: being on call 24/7 is a form of slavery, and anyone so busy and important should have a personal assistant instead. A similar reasoning could be applied to other wearable devices, including Google Glass. The truth is rather that being a sort of cyborg is not what people will embrace, but what they will try to avoid, unless it is inevitable. If this is not clear, consider current attempts to eliminate screens in favour of bodily projections, so that you may dial a telephone number by using a virtual keyboard appearing on the palm of your hand. This is a realistic scenario, but it is not what I mean by referring by the development of inforgs. Imagine instead the current experience of dialling a number by merely vocalizing it because your phone 'understands' you. You and your phone now share the same environment as two informational agents.

Second, when interpreting ourselves as informational organisms, I am not referring to the widespread phenomenon of 'mental outsourcing' and integration with our daily technologies. Of course, we are increasingly dependent on a variety of devices for our daily tasks, and this is interesting. However, the view according to which devices, tools, and other environmental supports or props may be enrolled as proper parts of our 'extended minds' is outdated. It is still based on a Cartesian agent, stand-alone and fully in charge of the cognitive environment, which is controlling and using through its mental prostheses, from paper and pencil to a smartphone, from a diary to a tablet, from a knot in the handkerchief to a computer.

Finally, I am not referring to a genetically modified humanity, in charge of its informational DNA and hence of its future embodiments. This post-humanism, once purged of its most fanciful and fictional claims, is something that we may see in the future, but it is not here yet, either technically (safely doable) or ethically (morally acceptable). It is a futuristic perspective.

What I have in mind is rather a quieter, less sensational, and yet more crucial and profound change in our conception of what it means to be human. We are regularly outsmarted and outperformed by our ICTs. They 'reckon' better than we do. And because of this, they are modifying or creating the environment in which we live. We have begun to understand ourselves as *inforgs* not through some biotechnological transformations in our bodies, but, more seriously and realistically, through the radical transformation of our environment and the agents operating within it. As we shall see in Chapter 7, in many contexts ICTs have already begun playing as the 'home' team in the infosphere with us as the 'away' team.

Enhancing, augmenting, and re-engineering technologies

The fourth revolution has brought to light the intrinsically informational nature of human identity. It is humbling, because we share such a nature with some of the smartest of our own artefacts. Whatever defines us uniquely, it can no longer be playing chess, checking the spelling of a document or translating it into another language, calculating the orbit of a satellite, parking a car, or landing an aircraft better than some ICTs. You cannot beat ICT even at a random game such as rock-paper-scissors, because the robot is so fast that, in 1 millisecond, it recognizes the shape that your hand is making, chooses the winning move, and completes it almost simultaneously. If you did not know better, you would think it was reading your mind.

The fourth revolution is also enlightening, because it enables us to understand ourselves better, as a special kind of informational organism. This is not equivalent to saying that we have digital alter egos, some Messrs Hydes represented by their @s, blogs, tweets, or https. This trivial point only encourages us to mistake ICTs for merely enhancing technologies, with us still at the centre of the infosphere. Our informational nature should not be confused with a 'data shadow' either, an otherwise useful term introduced to describe a digital profile generated from data concerning a user's habits online. The change is deeper. To understand it, consider the distinction between *enhancing* and *augmenting* technologies.

The handles, switches, or dials of enhancing technologies, such as axes, guns, and drills, are interfaces meant to plug the appliance into the user's body ergonomically. This is akin to the cyborg idea. Instead, the data and control panels of augmenting technologies are interfaces between different environments. On the one hand, there is the human user's outer environment. On the other hand, there is the environment of the technology. Some examples are the dynamic, watery, soapy, hot, and dark environment of the dishwasher; or the equally watery, soapy, hot, and dark but also spinning environment of the washing machine; or the still, aseptic, soapless, cold, and potentially luminous environment of the refrigerator. These technologies can be successful because they have their environments 'wrapped' and tailored around their capacities. This is the phenomenon of 'enveloping the world' that I shall analyse in Chapter 7. Now, despite some superficial appearances, ICTs are not merely enhancing or augmenting technologies in the sense just explained. They are forces that change the essence of our world because they create and re-engineer whole realities that the user is then enabled to inhabit. Their digital interfaces act as (often friendly) gateways. Let me give you an example.

Looking at the history of the mouse, one discovers that our technology has not only adapted to, but has also educated, us as users. Douglas Engelbart (1925–2013), the inventor of the mouse, once told me that he had experimented with a mouse to be placed under the desk, to be operated with one's knee, in order to leave the user's hands free. After all, we were coming from a past in which typewriters could be used more successfully by relying on both hands. Luckily, the story of the mouse did not go the same way as the story of the QWERTY keyboard, which never overcame the initial constraints imposed by the old typewriters.[9] Today, we expect to be able to touch the screen directly. Human–computer interaction is a symmetric relation.

To return to the initial distinction, the interface of a dishwasher is a panel through which the machine enters into the user's environment, while the interface of an ICT is a gate through which a user enters and can be present[10] in a region of the infosphere. This fundamental property of creating and opening new spaces underpins the many spatial metaphors of 'cyberspace', 'virtual reality', 'being online', 'surfing the Web', 'gateway', and so forth.

We are witnessing an epochal, unprecedented migration of humanity from its Newtonian, physical space to the infosphere itself as its new environment, not least because the latter is absorbing the former. As digital immigrants, like Generation X and Generation Y, are replaced by digital natives, like Generation Z, the latter will come to recognize no fundamental difference between the infosphere and the physical world, only a change in perspective. When the migration is complete, my guess is that Generation Z will increasingly feel deprived, excluded, handicapped, or poor to the point of paralysis and psychological trauma whenever it is disconnected from the infosphere, like fish out of water. One day, being an inforg will be so natural that any disruption in our normal flow of information will make us sick.

Conclusion

In light of the fourth revolution, we understand ourselves as informational organisms among others. We saw in Chapter 2 that, in the long run, *de-individualized* (you become 'a kind of') and *reidentified* (you are seen as a specific crossing point of many 'kinds of') inforgs may be treated like commodities that can be sold and bought on the advertisement market. We may become like Gogol's *dead souls*, but with wallets.[11] Our value depends on our purchasing power as members of a customer set, and the latter is only a click away. This is all very egalitarian: nobody cares who you are on the Web, as long as your ID is that of the right kind of shopper.

There is no stock exchange for these dead souls online, but plenty of Chichikovs (the main character in Gogol's novel) who wish to buy them. So what is the dollar of an inforg worth? As usual, if you buy them in large quantities you get a discount. So let's have a look at the wholesale market. In 2007, Fox Interactive Media signed a deal with Google to install the famous search engine (and ancillary advertising system) across its network of Internet sites, including the highly popular (at the time) MySpace. Cost of the operation: $900 million.[12] Estimated number of user profiles in MySpace: nearly 100 million at the time. So, average value of a digital soul: $9 at most, but only if it fitted the high-quality profile of a MySpace.com user. As Sobakievich, one of the characters in Gogol's novel, would say:

> It's cheap at the price. A rogue would cheat you, sell you some worthless rubbish instead of souls, but mine are as juicy as ripe nuts, all picked, they are all either craftsmen or sturdy peasants.[13]

The 'ripe nuts' are what really count, and, in MySpace, they were simply self-picked: tens of millions of educated people, with enough time on their hands (they would not be there otherwise), sufficiently well-off, English-speaking, with credit cards and addresses in deliverable places . . . it makes any advertiser salivate. Fast-forward five years. The market is bigger, the nuts are less ripe, and so the prices are even lower. In 2012, Facebook filed for a $5 billion initial public offering.[14] Divide that by its approximately 1 billion users at that time, and you have a price of $5 per digital soul. An almost 50 per-cent discount, yet still rather expensive. Consider that, according to the *Financial Times*,[15] in 2013 most people's profile information (an aggregate of age, gender, employment history, personal ailments, credit scores, income details, shopping history, locations, entertainment choices, address, and so forth) sold for less than $1 in total per person. For example, income details and shopping histories sold for $0.001 each. The price of a single record drops even further for bulk buyers. When I ran the online calculator offered by the *Financial Times*, the simulation indicated that 'marketers would pay approximately for your data: $0.3723'.

As a digital soul, in 2013, I was worth about a third of the price of a song on iTunes. You can imagine my surprise when, in 2013, Yahoo bought Tumblr (a blogging platform) for $1.1 billion: with 100 million users, that was, $11 per digital soul. I suspect it might have been overpriced.[16]

From Gogol to Google, a *personalizing*—recall the person who put a sticker in the window of her car, at the end of Chapter 2—reaction to such massive *customization* is natural, but also tricky. We saw that we could construct, self-brand, and reappropriate ourselves in the infosphere by using blogs and Facebook entries, Google homepages, YouTube videos, and Flickr albums; by sharing choices of food, shoes, pets, places we visit or like, types of holidays we take and cars we drive, instagrams, and so forth; by rating and ranking anything and everything we click into. It is perfectly reasonable that Second Life should be a paradise for fashion enthusiasts of all kinds. Not only does it provide a flexible platform for designers and creative artists, it is also the right context in which digital souls (avatars) intensely feel the pressure to obtain visible signs of self-identity. After all, your free avatar looks like anybody else. Years after the launch of Second Life, there is still no inconsistency between a society so concerned about privacy rights and the success of social services such as Facebook. We use and expose information about ourselves to become less informationally anonymous and indiscernible. We wish to maintain a high level of informational privacy almost as if that were the only way of saving a precious capital that can then be publicly invested (squandered, pessimists would say) by us in order to construct ourselves as individuals easily discernible and uniquely reidentifiable. Never before has informational privacy played such a crucial role in the lives of millions of people. It is one of the defining issues of our age. The time has come to have a closer look at what we actually mean by privacy after the fourth revolution.

5

PRIVACY

Informational Friction

The dearest of our possessions

'One of these days d'you think you'll be able to see things at the end of the telephone?' Peggy said, getting up.

S he will not return to her question again, in the remaining pages of Virginia Woolf's *The Years*.[1] The novel was published in 1937. Only a year earlier, the BBC had launched the world's first public television service in London, and Turing had published his ground-breaking work on computing machines.[2] Things were going to change dramatically.

Distracted by a technology that invites practical usage more readily than critical reflection, Peggy only half-perceives that ICTs are transforming society profoundly and irrevocably. The foundations of our information society were being laid down in the thirties. It was difficult to make complete sense of such a significant change in human history, at this early stage of its development. Today, the commodification of digital ICTs begun in the seventies, and the consequent spread of a global information society since the eighties, are progressively challenging the right to informational privacy, at least as Westerners still conceived it in Virginia Woolf's times. As inforgs inhabiting the infosphere, we are getting used to information flows being pervasive and respecting no boundaries. And yet, as Woolf wrote in an essay on Montaigne that she published in *The Common Reader* in 1925:[3]

> We, [...] have a private life [and] hold it infinitely the dearest of our possessions.

Today, we find protecting that dearest possession ever more difficult, in a social environment increasingly dependent on Peggy's futuristic technology.

The problem is pressing. It has prompted a stream of scholarly and scientific investigations, and there has been no shortage of political decisions and legally enforceable measures to tackle it. The ethical problem of privacy has become one of the defining issues of our hyperhistorical time. Browse textbooks in moral philosophy only a few decades old and you will find little or no reference to it. The goal of this chapter is not to review the extensive body of literature dedicated to informational privacy and its legal protection. Rather, it is to argue in favour of an interpretation of informational privacy as self-constituting, an interpretation that is coherent with, and complements, the facts and ideas presented in the previous chapters. In short, our task in this chapter is to understand informational privacy after the fourth revolution.

Privacies as freedoms from

It is common to distinguish four kinds of privacy. They can all be formulated in terms of 'freedoms from'. Let me quickly introduce them in no particular order of importance. First, there is Alice's *physical privacy*. This is her freedom from sensory interference or intrusion, achieved thanks to a restriction on others' ability to have bodily interactions with her or invade her personal space. Second, there is Alice's *mental privacy*. This refers to her freedom from psychological interference or intrusion, achieved thanks to a restriction on others' ability to access and manipulate her mental life. Third, there is Alice's *decisional privacy*. This is her freedom from procedural interference or intrusion, achieved thanks to the exclusion of others from decisions—especially but not only those concerning education, health care, career, work, marriage, and faith—taken by her and her group of intimates. And finally, there is Alice's *informational privacy*. This is her freedom from informational interference or intrusion, achieved thanks to a restriction on facts about her that are unknown or unknowable.

These four forms of privacy are often intertwined but they should not be confused. For the sake of simplicity, I shall treat them separately because in this chapter I shall deal only with the informational kind. So, by privacy, I shall mean informational privacy.

Two questions are going to lead our foray. Why have ICTs made privacy one of the most obvious and pressing issues in our society? And what is privacy after the fourth revolution? The answer to the second question will have to wait for the answer to the first, which in turn must wait until we have a better understanding of a concept that I briefly introduced in Chapter 1, that of 'informational friction'.

Informational friction

Informational friction refers to the forces that oppose the flow of information within a region of the infosphere. It is connected with the amount of effort required for some agent to obtain, filter, or block information about other agents in a given environment, by decreasing, shaping, or increasing informational friction. To see how, consider the following scenario.

Four students, Alice, Bob, Carol, and Dave, our inforgs, live in the same house, a region of the infosphere. Intuitively, the larger the *informational gap* among them, the less they know about each other, and the more private their lives can be. The informational gap depends on the degree of *accessibility* of their personal information. In our example, there will be more or less privacy depending on whether the students have their own rooms and en suite bathrooms, for example. Accessibility, in turn, depends on the nature of the inforgs, the environment in which they are embedded, and the interactions that can take place in that environment by those inforgs. If the walls in the house are few and thin and all the students have excellent hearing, then the degree of accessibility is increased, the informational gap is reduced, and privacy is more difficult to protect. The love lives of the students may be deeply affected by the Japanese-style house they have chosen to share. Or suppose that all the walls and the furniture in our students' house are transformed into perfectly transparent glass and all our students have perfect sight. In this space resembling Bentham's *Panopticon*[4] (a place so designed that it is entirely visible from a single point) any privacy will become virtually impossible. Consider, finally, a science-fiction scenario regarding time. In *The Dead Past*,[5] Asimov describes a *chronoscope*, a device that allows direct observation of past events. The chronoscope turns out to be of only limited use for archaeologists, since it can look only a couple of centuries into the past. However, people soon discover that it can easily be tuned into the most recent past, with a time lag of fractions of seconds, that is, with little informational friction. If our students could use Asimov's chronoscope they could monitor any event almost in real time. This too would be the end of privacy, for the dead past is only a synonym for 'the living present', as one of the characters in Asimov's story remarks rather philosophically. Clearly, the informational affordances and constraints provided by an environment are such only in relation to agents with specific informational capacities. The debate[6] on privacy issues in connection with the design of office spaces—from private offices to panel-based open-plan office systems, to completely open working environments—offers a significant example of the relevance of varying degrees of informational friction in social contexts. Next time you draw the curtains in your living room, you know you are increasing the informational friction in the environment.

We are now ready to formulate a qualitative kind of equation. Given some amount of personal information available in a region of the infosphere, the lower the informational friction in that region, the higher the accessibility of personal information about the agents embedded in that region, the smaller the informational gap among them, and the lower the level of privacy that can be expected. Put simply and more generally, *privacy is a function of the informational friction in the infosphere*. Any factor decreasing or increasing informational friction will also affect privacy. So it may seem that we have an answer to our first question. ICTs have made privacy one of the most obvious and pressing issues in our society because they unquestionably and influentially affect informational friction. Unfortunately, things are a bit more complicated. The previous answer is a good stepping stone, but it still fails to account for two important phenomena, without which it is difficult to answer our second question. Each of them deserves a specific section, but let me first outline them here briefly. First, although ICTs may erode informational friction, anonymity may counterbalance their impact. And second, old ICTs, such as the radio and the TV, affect informational friction only one way, that is, by decreasing it, whereas new ICTs work both ways, that is, they can decrease or increase it, so they can reduce or enhance the degree of privacy we enjoy. The next two sections are dedicated to these two phenomena. At the end of them, we will finally revise our answer to our first question and be ready to move to the second one.

Anonymity

During the nineteenth and the twentieth centuries, while old ICTs such as the telegraph, the radio, photography, the telephone, and the TV were progressively erasing informational friction, the social phenomenon of the new metropolis counteracted their effects. Urban environments fostered a type of privacy based on *anonymity*. Anonymity may be understood as the unavailability of personal information, due to the difficulty of collecting or correlating different bits of information about someone. This is the sort of privacy enjoyed by a leaf in the forest, still inconceivable nowadays in rural settings or small villages, where everyone tends to know everyone else.

Anonymity made modern societies enjoy an unprecedented degree of privacy, even if by proxy, despite the growing decrease of informational friction caused by the development of old ICTs. In their classic article 'The Right to Privacy', published in the *Harvard Law Review* in 1890, Samuel D. Warren and Louis Brandeis warned that privacy was being undermined by

> recent inventions and business methods [...], instantaneous photographs and newspaper enterprise [...] and numerous mechanical devices.[7]

And yet, the power of such technologies was counterbalanced by countervailing forces. Stevenson's *The Strange Case of Dr Jekyll and Mr Hyde* was first published in 1886. A year later, Conan Doyle published *A Study in Scarlet*. In the same period in which Warren and Brandeis were working on their classic article, the Edinburgh of Dr Jekyll and the London of Sherlock Holmes already provided increasing opportunities for privacy through anonymity, despite the recent availability of new technologies. Sometimes, it seems that the privacy we miss nowadays is really nineteenth-century anonymity.

Because digital ICTs are modifying our informational environments, our interactions, and ourselves it would be naïve to expect that privacy in the future will mean exactly what it meant in the industrial Western world in the middle of the last century.[8] The information society has revised the threshold of informational friction and therefore provides a different sense in which its citizens appreciate their privacy. In a way, a different kind of privacy is the price we pay to enter into hyperhistory. Society cannot depend so widely and deeply on ICTs without allowing them to reshape the environment and what happens within it. There is already a significant difference in how Generation X and Generation Y perceive privacy. A report by the Pew Internet & American Life Project on 'Teens, Privacy and Online Social Networks' indicates that,

> To teens, all personal information is not created equal. They say it is very important to understand the context of an information-sharing encounter.[9]

And a more recent report by the Berkman Center for Internet & Society rightly stresses that,

> for youth, 'privacy' is not a singular variable. Different types of information are seen as more or less private; choosing what to conceal or reveal is an intense and ongoing process [...]. Rather than viewing a distinct division between 'private' and 'public,' young people view social contexts as multiple and overlapping. [...] Indeed, the very distinction between 'public' and 'private' is problematic for many young people, who tend to view privacy in more nuanced ways, conceptualizing Internet spaces as 'semi-public' or making distinctions between different groups of 'friends' [...]. In many studies of young people and privacy, 'privacy' is undefined or is taken to be an automatic good. However, disclosing information is not *necessarily* risky or problematic; it has many social benefits that typically go unmentioned.[10]

Generation Z will probably move even further away from what current middle-age academics implicitly consider an obvious and indisputable sense of informational privacy, since it is already growing up in an infosphere of double negation, of *a-anonymity*.

We have seen that, at the end of the nineteenth century, the informational friction in the infosphere, actually reduced by old ICTs, was nevertheless increased by social conditions favouring anonymity, and hence a new form of privacy as anonymity. In this respect, the diffusion of new ICTs has finally brought to completion the process that began with the invention of printing. We are now back into the digital community, where anonymity can no longer be taken for granted, and hence where the decrease in informational friction caused by old and new ICTs can demonstrate its full-blown effects on privacy. In the UK, for example, the digital ICTs that allowed terrorists to communicate undisturbed over the Internet were also responsible for the identification of the London bombers in a matter of hours in 2005. Sadly, the same happened again with the Boston bombing in 2013. Likewise, mobile phones are increasingly useful as forensic evidence in trials. In the UK, cell site analysis (a form of triangulation that estimates the location of a mobile phone when it is used) helped disprove Ian Huntley's alibi and convict him for the murders of Holly Wells and Jessica Chapman. Sherlock Holmes has the means to fight Mr Hyde.

Your supermarket knows exactly what you like, but so did the owner of the grocery store where your grandparents used to shop. Your bank has detailed records of all your visits and of your financial situation, but how exactly is this different from the old service? A phone company could analyse and transform the call data collected for billing purposes into a detailed subscriber profile: social network (names and addresses of colleagues, friends, or relatives called), possible nationality (types of international calls), times when one is likely to be at home and hence working patterns, financial profile (expenditure), and so forth. Put together the data from the supermarket, the bank, and the phone company, and inferences of all sorts could be drawn for your credit rating. Yet so they could be, and were, in Alexandre Dumas's *The Count of Monte Cristo* (1844). *Some* steps forward into the information society are really steps back into a small community and, admittedly, the claustrophobic atmosphere that may characterize it.

How difficult is it to live in a glass infosphere? Human agents tend to be acquainted with different environments that have varying degrees of informational friction and hence to be rather good at adapting themselves accordingly. As with other forms of fine balances, it is hard to identify, for all agents in any environments and circumstances, a common, lowest threshold of informational friction below which human life becomes increasingly unpleasant and ultimately unbearable, although perhaps Orwell has described it well in *Nineteen Eighty-Four*.[11] It is clear, however, that a particular threshold has been reached when agents are willing to employ resources, run risks, or expend energy to restore it, for example by building a higher fence, by renouncing a desired service, or by investing time in revising a customer profile. Different agents have different degrees of sensitivity. One needs to remember that several factors (character, culture, upbringing, past experiences, etc.) make each of us a unique individual. To one person, a neighbour capable of seeing one's garbage in the garden may seem an unbearable breach of their privacy, which it is worth any expenditure and effort to restore; to another person, living in the same room with several other family members may feel entirely unproblematic. Human agents can adapt to very low levels of informational friction. Virginia Woolf's essay on Montaigne discusses the lack of informational friction that characterizes public figures in public contexts, an issue that reacquired all its pregnancy in the UK because of the phone-hacking scandal that led to the closure of the *News of the World*. Politicians and actors are used to environments where privacy is a rare commodity and informational friction assumed to be non-existent. Likewise, people involved in 'Big Brother'-style (although 'Truman Show' would be a more appropriate label) programmes show a remarkable capacity to adapt to settings where any informational friction between them and the public is systematically reduced, apparently in the name of entertainment. In far more tragic and realistic contexts, prisoners in concentration camps are subject to extreme duress due to both intended and unavoidable decrease of informational friction.

In the early history of the Web, roughly when Netscape was synonymous with browser, users believed that being online meant being entirely anonymous. Actions lost their named sources, and untraceability felt like privacy. A networked computer was like Gyges' ring, the magical artefact that enables its owner to become invisible at will. Plato used it in his *Republic* to discuss what an ordinary person would do if he could act without any fear of being caught and punished. He was not optimistic:

> Suppose now that there were two such magic rings, and the just put on one of them and the unjust the other; no man can be imagined to be of such an iron nature that he would stand fast in justice. No man would keep his hands off what was not his own when he could safely take what he liked out of the market, or go into houses and lie with anyone at his pleasure, or kill or release from prison whom he would, and in all respects be like a god among men.[12]

We basically ran Plato's social experiment online for some years. The result was close to what Plato suspected: Internet users behaved potentially less responsibly, socially speaking. Things have changed. Turing would certainly have appreciated Peter Steiner's cartoon discussed in Chapter 3, in which two dogs boast about their anonymity. We saw that nowadays it is no longer funny, only outdated. Cookies,[13] monitoring software, and malware (malicious software, such as spyware) have made more and more of us realize that the screen in front of us is not a shield for our privacy or Harry Potter's invisibility cloak, but a window onto our lives online, through which virtually anything may be seen. There is no magic formula. Controversial technologies, labelled 'respawning', can reproduce tracking cookies even after a user has specifically deleted them.[14] We expect websites to monitor and record our activities and do not even seem to mind for what purpose. It is not that we do not care about privacy, but that we accept that being online may be one of the less private things in our life. The screen is a monitor and is monitoring you. 'You are being watched' 24/7 by 'the machine', as Harold Finch, the reclusive billionaire computer genius reminds us at the beginning of every episode of *Person of Interest*, CBS television's crime drama series.

In 1999, a journalist at *The Economist* ran an experiment still worth reporting today.[15] He asked a private investigator, 'Sam', to show what information it was possible to gather about someone. The journalist himself was to be the subject of the experiment. The country was the UK, the place where the journalist lived. The journalist provided Sam with only his first and last names. Sam was told not to use

> any real skulduggery (surveillance, going through the domestic rubbish, phone-tapping, hacking, that sort of thing).

The conclusion? By using several databases and various ICTs,

> Without even talking to anyone who knows me, Sam [...] had found out quite a bit about me. He had a reasonable idea of my personal finances— the value of my house, my salary and the amount outstanding on my mortgage. He knew my address, my phone number, my partner's name, a

former partner's name, my mother's name and address, and the names of three other people who had lived in my house. He had 'found' my employer. He also had the names and addresses of four people who had been directors of a company with me. He knew my neighbours' names.

Shocking? Yes, in the anonymous industrial society of modern times, but not really in the pre-industrial village that preceded it, or in the hyperhistorical, information society that comes after it. In Guarcino, a small village south of Rome of roughly 1,000 people, everybody knows everything about everybody else, 'vita, morte e miracoli', 'life, death, and miracles', as they say in Italy. Informational friction is very low, anonymity does not redress the balance, and so there is precious little privacy.

There are, of course, many dissimilarities between the small local village and the global digital one. History may repeat itself, yet never too monotonously. Small communities have a high degree of intra-community transparency (like a shared house) but a low degree of inter-community transparency (they are not like the Big Brother house, visible to outside viewers). So in those communities, breaches of privacy are reciprocal, yet there are few breaches of privacy across the boundary of the community. This is quite different from today's information society. There can be little transparency within the communities we live or work in (we hardly know our neighbours, and our fellow workers have their privacy rigorously protected), yet data-miners, hackers, and institutions can be well informed about us. There is no symmetry. Breaches of privacy from outside are common. What is more, we do not even know whether they know our business. Part of the value of the comparison between the past and the present lies in the size of the community taken into consideration. A special trait of the information society is precisely its lack of boundaries, its global nature. We live in a single infosphere, which has no 'outside' and where intra- and inter-community relations are more difficult to distinguish. The types of invasion of privacy are quite different too. In the small community, breaches of privacy may shame or discredit you. Interestingly, the philosopher and Church Father Augustine of Hippo (354–430) usually speaks of privacy in relation to the topic of sexual intercourse in married couples, and he always associates it with secrecy and then secrecy with shame or embarrassment. Or breaches of privacy may disclose your real identity or character. Things that are private became public knowledge. In the information society, such breaches involve the unauthorized collection of information, not necessarily its publication. Things that are private may not become public at all; they may be just accessed and used by privileged others. The small village is self-regulating and this limits breaches of privacy. Everyone knows that they are as subject to scrutiny as everyone else, and this sets an unspoken limit on their enthusiasm for intruding into others' affairs. There is not such social constraint in the global digital village. But other defences have become available. Today the information society has the digital means to protect what the small village must necessarily forfeit, as we shall see in the next section.

Empowerment

Earlier I promised to analyse two phenomena. We just considered how the decrease of informational friction, caused by ICTs, might be counterbalanced by other factors, especially modern anonymity. The second phenomenon, the topic of this section, concerns the difference between old and new ICTs.

Old ICTs have always tended to reduce what agents considered the normal degree of informational friction in their environment. This already held true for the invention of the alphabet or the diffusion of printing. We saw that in 1890 Warren and Brandeis complained that photography and the rise of the daily press further increased this trend. Tele-ICTs, from the telescope to the television, and recording ICTs, from the alphabet to the smartphone app, cannot but reduce the informational friction in the infosphere. *Rear Window*, the classic film directed by Alfred Hitchcock (1899–1980) in 1954, provides a splendid illustration. A journalist, L. B. 'Jeff' Jefferies (James Stewart), while confined to a wheelchair because of a broken leg, is still able to spy on his neighbours and solve a crime thanks to a variety of technologies. Twenty years later, the Watergate scandal and Nixon's resignation in 1974 owed as much to ICTs.

Those who control old pre-digital ICTs control the informational friction and hence the information flows. Such an empowering inter-pretation of ICTs is well represented by dystopian views of informa-tionally omnipotent agents, able to overcome any informational friction, to control every aspect of the information flow, to acquire any personal data, and hence to implement the ultimate surveillance system, thus destroying all privacy. The loss of 'the dearest of our possessions' is a pre-digital problem. Recall that Orwell's *Nineteen Eighty-Four*, first published in 1949, contains no reference to computers or digital machines.

Now, given this overall picture, it is understandable that a 'conti-nuist' interpretation of technological changes would suggest that new digital ICTs should be treated as just one more instance of the well-known enhancement or augmentation of old ICTs. But then—the reasoning goes—if there is no radical difference between old and new ICTs, it is also reasonable to argue that the latter cause increasing problems for privacy merely because they are orders of magnitude more powerful than past technologies in empowering agents in the infosphere. 'Big Brother', the character in *Nineteen Eighty-Four* by George Orwell (1903–50), is readily associated today with the ultimate database.

The trouble with this reasoning is that, contrary to old ICTs, new ICTs empower users in both directions, as they can both increase and decrease informational friction.

Empowerment comes in two main flavours. Both count for our present purposes. Empowerment may mean 'equal opportunities'. This is empowerment as *inclusion* in decision-making processes, as opposed to marginalization, exclusion, or discrimination. It is what we have in mind when talking about gender or minority empower-ment. In a decent democracy, this kind of empowerment is or should soon become unnecessary. Then there is the 'more opportunities' sense. This is empowerment as *improvement* in the quantity and quality of available choices. It is the sort of empowerment in question when we discuss consumers' experiences or interactions, for example. Even in an ideal democracy, there is no limit to how far this second empowerment should go, because there is no limit to the nature and number of opportunities that may be provided, not least because the latter are a matter of human development. Now, both forms of empowerment are increasingly linked to available and accessible information. Both are needed in order to ensure more equality and better standards of living. And, in some cases, both are joined in a single sense, as when patients' empowerment is in question, or, as I shall argue presently, the empowerment of inforgs by ICTs.

In the infosphere, we as inforgs are increasingly empowered (more inclusion and more improvement) by new ICTs not only to gather and process personal data, but also to control and protect them. Recall that the digital now deals effortlessly with the digital. The phenomenon cuts both ways. It has led not only to a huge expansion in the flow of personal information being recorded, processed, and exploited, but also to a large increase in the types and levels of control that agents can exercise on their personal data. For example, reputation management companies that monitor and improve information about an individual or brand online are mushrooming. In 2013, one of them, Reputation.com, had over 1 million clients, in over 100 countries. And while there is only a certain amount of personal information that one may care to protect, the potential growth of digital means and measures to control its life cycle does not seem to have a foreseeable limit. Suppose privacy is the right of individuals (be these single persons, groups, or institutions) to control the life cycle (especially the generation, access, recording, and usage) of their information and determine when, how, and to what extent their information is processed by others. Then one must agree that digital ICTs may enhance as well as hinder the possibility of enforcing such a right.

At the point of *data generation*, digital ICTs can foster the protection of personal data, especially by means of encryption, anonymization, password encoding, firewalling, specifically devised protocols or services, and, in the case of externally captured data, warning systems. At the point of *data storage*, digital ICTs make possible legislation, such as the Data Protection Directive already passed by the EU in 1995, which can guarantee that no informational friction, already removed by digital ICTs, is surreptitiously reintroduced to prevent agents from discovering the existence of personal data records, and from accessing them, checking their accuracy, correcting or upgrading them, or demanding their erasure. And at the point of *data management*— especially through data mining, sharing, matching, and merging— digital ICTs can help agents to control and regulate the usage of their data by facilitating the identification and regulation of the relevant users involved. At each of these three stages, solutions to the problem of protecting privacy can be not only self-regulatory and legislative but also technological, not least because privacy infringements can more easily be identified and redressed, also thanks to digital ICTs.

All this is not to say that we are inevitably moving towards an idyllic scenario in which our PETs (Privacy Enhancing Technologies) will fully protect our private lives and information against harmful PITs (Privacy Intruding Technologies). Such optimism is unjustified. Solutions will not develop by themselves without some effort on our part. But it does mean that digital ICTs are already providing some means to counterbalance the risks and challenges that they represent for privacy, and hence that no fatalistic pessimism is justified either. Digital ICTs do not necessarily erode privacy; they can also enhance and protect it. They may have eroded anonymity as a proxy for privacy, but they have introduced privacy through the proper design of our technologies and social environments.

We have come to the end of our two forays. As promised, we are now ready to revise our answer to our first question. New ICTs have made privacy one of the most obvious and pressing issues in our society not only because they have continued to erode informational frictions, like old ICTs did, but also because they have undermined a counterbalancing form of privacy based on anonymity and have empowered agents both ways, to decrease and increase informational friction.

The time has come to turn to the second question: what is privacy after the fourth revolution?

Why privacy matters

Two theories about the value of our privacy are particularly popular: the reductionist interpretation and the ownership-based interpretation.

The reductionist interpretation argues that the value of privacy rests on a variety of undesirable consequences that may be caused by its breach, either personally, such as distress, or socially, such as unfairness. Privacy is a utility, also in the sense of providing an essential condition of possibility of good human interactions, by preserving human dignity or by guaranteeing political checks and balances, for example.

The ownership-based interpretation argues that informational privacy needs to be respected because of each person's rights to bodily security and property, where 'property of x' is classically understood as the right to exclusive use of x. A person is said to *own* his or her information (information about him- or herself)—recall Virginia Woolf's 'infinitely the dearest of our *possessions*'—and therefore to be entitled to control its whole life cycle, from generation to erasure through usage.

The two interpretations are not incompatible, but they stress different aspects of the value of privacy. The reductionist interpretation is more oriented towards a consequentialist assessment of privacy in terms of cost-benefit analyses of its protection or violation. The ownership-based interpretation is more oriented towards a 'natural rights' understanding of the value of privacy itself, in terms of private or intellectual property. Unsurprisingly, because they both belong to a 'historical mentality', they both compare privacy breach to trespass or unauthorized invasion of, or intrusion in, a metaphorical space or sphere of personal information, whose accessibility and usage ought to be fully controlled by its owner and hence kept private.

Neither interpretation is entirely satisfactory. The reductionist interpretation defends the need for respect for privacy in view of the potential misuse of the information acquired. So it is certainly reasonable, especially from a consequentialist perspective. But it may be inconsistent with pursuing and furthering social interests and welfare. Although it is obvious that some public personal information may need to be protected—especially against profiling or unrestrained electronic surveillance—it remains unclear, on a purely reductionist basis, whether a society devoid of any privacy may not be a better society after all, with a higher common welfare. Indeed, it has been convincingly argued that the defence of privacy in the home may actually be used as a subterfuge to hide the dark side of privacy: domestic abuse, neglect, or mistreatment.[16] Precisely because of reductionist-only considerations, even in democratic societies we tend to acknowledge that the right to privacy can be overridden when other concerns and priorities, including public safety or national security, become more pressing. All this by putting some significant interpretative pressure on the 'arbitrary' clause that qualifies article 12 of *The Universal Declaration of Human Rights* which states that

> No one shall be subjected to *arbitrary* [emphasis added] interference with his privacy, family, home or correspondence, nor to attacks upon his honour and reputation. Everyone has the right to the protection of the law against such interference or attacks.

The ownership-based interpretation also falls short of being entirely satisfactory, for at least three reasons.

First, informational contamination may undermine passive informational privacy. This is the unwilling acquisition of information or data, including mere noise, imposed on someone by some external source. Brainwashing may not occur often, but junk mail, or the case of a person chatting loudly on a phone nearby, are unfortunately common experiences of passive privacy breach, yet no informational ownership seems to be violated.

Second, there is a problem of privacy in public contexts. Privacy is often exercised publicly, that is, in spaces that are socially, physically, and informationally shared: anyone can see what one is doing downtown. How could a CCTV system be a breach of someone's privacy if the person in question is accessing a space that is public in all possible senses anyway? The ownership-based interpretation cannot provide a satisfactory answer.

And finally, there is a metaphorical and imprecise use of the concept of 'information ownership', which cannot quite explain the lossless acquisition or usage of information. We saw in Chapter 2 that information is not like a pizza or a CD: contrary to other things that one owns, one's personal information is not lost when acquired by someone else. Analyses of privacy based on 'ownership' of an 'informational space' are metaphorical twice over. We need a better alternative, so here is a proposal.

The self-constitutive value of privacy

Both the reductionist and the ownership-based interpretation fail to acknowledge the significant changes brought about by digital ICTs. They belong to an industrial culture of material goods and of manufacturing/trading relations. They rely on conceptual frameworks that are more 'historical' than 'hyperhistorical', so they are overstretched when trying to cope with the new challenges offered by an informational culture of services and usability. Interestingly, Warren and Brandeis had already realized this limit with impressive insight:

> where the value of the production [of some information] is found not in the right to take the profits arising from publication, but in the peace of mind or the relief afforded by the ability to prevent any publication at all, it is difficult to regard the right as one of property, in the common acceptation of the term [emphasis added].[17]

More than a century later, in the same way that the information revolution is best understood as a fourth revolution in our self-understanding, privacy requires an equally radical reinterpretation, one that takes into account the informational nature of our selves and of our interactions as inforgs.

Such a reinterpretation is achieved by considering each person as constituted by his or her information, and hence by understanding a breach of one's informational privacy as a form of aggression towards one's personal identity. This interpretation of privacy as having a self-constituting value is consistent with the fact that ICTs can both erode and reinforce informational privacy, and hence that a positive effort needs to be made in order to support not only Privacy Enhancing Technologies but also constructive applications, which may allow users to design, shape, and maintain their identities as informational agents. The value of privacy is both to be defended and enhanced.

The information flow needs some friction in order to keep firm the distinction between the macro multi-agent system (the society) and the identity of the micro multi-agent systems (the individuals) constituting it. Any society (even a utopian one) in which no informational privacy is possible is one in which no self-constituting process can take place, no personal identity can be developed and maintained, and hence no welfare can be achieved, social welfare being only the sum of the individuals' involved. The total 'transparency' of the infosphere that may be advocated by some reductionists—recall the example of the house and of our students living inside it—achieves the protection of society only by erasing all personal identity and individuality, a 'final solution' for sure, but hardly one that the individuals themselves, constituting the society so protected, would be happy to embrace. As has been correctly remarked:

> the condition of no-privacy threatens not only to chill the expression of eccentric individuality, but also, gradually, to dampen the force of our aspirations to it.[18]

The advantage of the self-constituting interpretation over the reductionist one is that consequentialist concerns may override respect for privacy, whereas the self-constituting interpretation, by equating its protection to the protection of personal identity, considers it a fundamental right. By default, the presumption should always be in favour of its respect. As we shall see, this is not to say that privacy is never negotiable in any degree.

Looking at the nature of a person as being constituted by that person's information enables one to understand the right to privacy as a right to personal immunity from unknown, undesired, or unintentional changes in one's own identity as an informational entity, both *actively* and *passively*. Actively, because collecting, storing, reproducing, manipulating, etc. Alice's information amounts now to stages in stealing or cloning her personal identity. Passively, because breaching Alice's privacy may now consist in forcing her to acquire unwanted data, thus altering her nature as an informational entity without consent. Brainwashing is as much a privacy breach as mind-reading. The first difficulty facing the ownership-based interpretation is thus avoided. The self-constituting interpretation suggests that your informational sphere and your personal identity are co-referential, or two sides of the same coin. There is no difference because 'you are your information', so anything done to your information is done to you, not to your belongings. It follows that the right to privacy, both in the active and in the passive sense just seen, shields one's personal identity. This is why privacy is extremely valuable and ought to be respected.

The second problem affecting the ownership-based interpretation is also solved because violations of informational privacy are now more fruitfully compared to kidnapping rather than trespassing. The advantage, in this change of perspective, is that it becomes possible to dispose of the false dichotomy qualifying privacy in public or in private contexts. Some information constitutes Alice context-independently, and therefore Alice is perfectly justified in wishing to preserve her integrity and uniqueness even in entirely public places.

Trespassing makes no sense in a public space, but kidnapping is a crime independently of where it is committed.

As for the third problem, one may still argue that an agent 'owns' his or her information, yet no longer in the metaphorical sense just seen, but in the precise sense in which an agent *is* her or his information. 'Your' in 'your information' is not the same 'your' as in 'your car' but rather the same 'your' as in 'your body', 'your feelings', 'your memories', 'your ideas', 'your choices', and so forth. It expresses a sense of constitutive *belonging*, not of external *ownership*, a sense in which your body, your feelings, and your information are part of you but are not your (legal) possessions. Once again, it is worth quoting Warren and Brandeis, this time at length:

> the protection afforded to thoughts, sentiments, and emotions [...] is merely an instance of the enforcement of the more general right of the individual to be let alone. It is like the right not to be assaulted or beaten, the right not to be imprisoned, the right not to be maliciously persecuted, the right not to be defamed [or, the right not to be kidnapped, my addition]. In each of these rights [...] there inheres the quality of being owned or possessed and [...] there may be some propriety in speaking of those rights as property. But, obviously, they bear little resemblance to what is ordinarily comprehended under that term. *The principle [...] is in reality not the principle of private propriety but that of inviolate personality* [emphasis added].[19] [...] *the right to privacy, as part of the more general right to the immunity of the person, [is] the right to one's personality* [emphasis added].[20]

This self-constituting conception of privacy and its value has started being appreciated by more advanced hyperhistorical societies, in which identity theft is one of the fastest growing offences. Privacy is the other side of identity theft; to the point that, ironically, for every person whose identity has been stolen (around 10 million Americans are victims annually) there is another person (the thief) whose identity has been 'enhanced'.

Problems affecting companies such as Google or Facebook and their privacy policies convey a similar picture. As Kevin Bankston, staff attorney at the Electronic Frontier Foundation, once remarked:[21]

Your search history shows your associations, beliefs, perhaps your medical problems. The things you Google for *define* you [emphasis added]. [...] data that's practically a printout of what's going on in your brain: What you are thinking of buying, who[m] you talk to, what you talk about.

The questions you ask, what you are looking for, identify you better than the answers you give, for they can lie so much less.

As anticipated, the self-constituting interpretation reshapes some of the assumptions behind a still 'industrial', 'modern', or 'Newtonian' conception of privacy. The following considerations illustrate such a transition.

If personal information is finally acknowledged to be a constitutive part of someone's personal identity and individuality, then one day it may become strictly illegal to trade in some kinds of personal information, exactly as it is illegal to trade in human organs (including one's own) or slaves. The problems of pornography and violence may also be revisited in the light of a self-constituting interpretation of privacy. Whatever you are exposed to runs the risk of ending up constituting you. Think of it as food that will be absorbed by your body and become part of you. If you are not careful, if you have no defences, an early exposure may be lethal or injure you for ever. How many things are there that you wish you had never seen, or been told, or heard? We must protect children's privacy exactly because ICTs are technologies that shape the self. At the same time, we might relax our attitude towards some kinds of 'dead personal information' that, like 'dead pieces of oneself', are not really, or no longer, constitutive of ourselves. Legally, Alice may not sell her kidney, but she may sell her hair or be rewarded for giving blood. Recall the experiment of the journalist at *The Economist*. Little of what Sam had discovered could be considered constitutive of the person in question. We are constantly leaving behind a trail of data, pretty much in the same sense in which we are shedding a huge trail of dead cells. The fact that nowadays digital ICTs allow our data trails to be recorded, monitored, processed, and used for social, political, or commercial purposes is a strong reminder of our informational nature as individuals. It might be seen as a new level of environmentalism, as an increase in what is recycled and a decrease in what is wasted. At the moment, all this is just speculation and in the future it will probably be a matter of fine adjustments of ethical sensibilities, but the Third Geneva Convention (1949) already provides a clear test of what might be considered 'dead personal information'. A prisoner of war need only give his or her name, rank, date of birth, and serial number, and no form of coercion may be inflicted on him or her to secure any further information of any kind. If we were all treated fairly as 'prisoners of the information society', our privacy would be well protected and yet there would still be some personal data that would be perfectly fine to share with any other agent, even hostile ones. It is not a binary question of all or nothing, but an analogue one of fine balance and degree.

A further issue that might be illuminated by looking at privacy from a self-constituting perspective is that of confidentiality. The sharing of private information with someone, implicitly, especially by doing things together, or explicitly, is based on a relation of profound trust that binds the agents involved intimately. This coupling is achieved by allowing the agents to be partly constituted as selves by the same information. Visually, the informational identities of the agents involved now overlap, at least partially. The union of the agents forms a single unity, a supra-agent, or a new multi-agent individual. Precisely because entering into a new supra-agent is a delicate and risky operation, care should be exercised before 'melding' oneself with other individuals by sharing personal information or its source, such as common experiences. This is the way I interpret the concluding sentence of *The Catcher in the Rye*, the famous novel by J. D. Salinger (1919–2010):

Don't tell anybody anything. If you do, you start missing everybody.[22]

Confidentiality is an intimate bond that is hard and slow to forge properly, yet resilient to many *external* forces when finally in place, as the supra-agent is stronger than the constitutive agents themselves.

Relatives, friends, classmates, fellows, colleagues, comrades, compan-ions, partners, teammates, spouses, and so forth may all have experi-enced the nature of such a bond, the stronger taste of a 'we'. But it is also a bond brittle and difficult to restore when it comes to *internal* betrayal, since the disclosure, deliberate or unintentional, of some personal information in violation of confidence can entirely and irrecoverably destroy the privacy of the new supra-agent born out of the joining agents, by painfully tearing them apart. The 'we' is strongly armoured against 'the other', but extremely fragile against the internal betrayal from 'one of us'.

A final issue can be touched upon rather briefly, as it has already been mentioned: the self-constituting interpretation stresses that priv-acy is also a matter of construction of one's own identity. Your right to be left alone is also your right to be allowed to experiment with your own life, to start again, without having records that mummify your personal identity for ever, taking away from you the power to form and mould who you are and can be. Every day, a person may wish to build a different, possibly better, 'I'. We never stop becoming our-selves, so protecting a person's privacy also means allowing that person the freedom to construct and change herself profoundly. The right to privacy is also the right to a renewable identity.

Biometrics

On 12 September 1560 the young Montaigne attended the public trial of Arnaud du Tilh, an impostor who was sentenced to death for having faked his identity. Many acquaintances and family members, including his wife Bertrande, had apparently been convinced for a long while that he was Martin Guerre, returned home after many years of absence. Only when the real Martin Guerre showed up was Ar-naud's actual identity finally ascertained.

Had Martin Guerre always been able to protect his personal infor-mation, Arnaud du Tilh would have been unable to steal his identity. Clearly, the more one's privacy is protected, the more one's personal identity can be safeguarded. This new qualitative equation is a direct consequence of the self-constituting interpretation. Personal identity also depends on informational privacy. The difficulty facing our contemporary society is how to combine the new equation with the other equation introduced earlier, according to which informational privacy is a function of the informational friction in the infosphere. Ideally, one would like to reap all the benefits from

1. the highest level of information flow; and hence from
2. the lowest level of informational friction; while enjoying
3. the highest level of informational privacy protection; and hence
4. the highest level of personal identity protection.

The problem is that (1) and (4) seem incompatible. If you facilitate and increase the information flow through digital ICTs, then the protection of one's personal identity is bound to come under increasing pressure. You cannot have an identity without having an identikit.

The problem starts looking less daunting once we realize an important difference. The information flowing in (1) consists of all sorts of data, including *arbitrary* data *about* oneself (like a name and surname, National Insurance Number, etc.) that are actually shareable without any self-constituting harm. Recall what the Geneva Convention prescribes in the case of information that can be secured from a prisoner of war. However, the information required to protect (4) refers to *constitutive* data, that is, data *that make you yourself*, such as your intimate beliefs, or your unique emotional involvement. These are the sorts of data that need to be safeguarded in order to protect the individual that embodies them. The distinction becomes clearer and more pressing when discussing privacy and biometrics, as we shall see in this section.

Personal identity is the weakest link and the most delicate element in our problem. Even nowadays, personal identity is regularly protected and authenticated by means of some *arbitrary* data, *randomly* or *conventionally* attached to the bearer/user, like a mere label: a name, an address, a Social Security number, a bank account, a credit card number, a driving licence number, a PIN, and so forth. None of these bits of information constitutes you. Each label in the list has no intimate link with its bearer; it is merely associated with your identity and can easily be detached from it without affecting your self. The rest is a mere consequence of this 'detachability'. The more the informational friction in the infosphere decreases, the swifter these detached labels can flow around, and the easier it becomes to grab and steal them and use them for illegal purposes. Arnaud du Tilh had stolen a name and a profile and succeeded in impersonating Martin Guerre for many years in a rather small village, within a community that knew him well, fooling even Martin's wife (apparently). Eliminate all personal interactions, promote a culture of proxies, and identity theft becomes the easiest thing in the world.

A quick and dirty way to fix the problem would be to clog the infosphere by slowing down the information flow; building some traffic-calming devices, as it were. This seems to be the sort of policy popular among some IT officers and middle-ranking bureaucrats, keen on not allowing this-or-that operation for security reasons. However, as with all counter-revolutionary or anti-historical (anti-hyperhistorical, to be precise) approaches, 'resistance is futile'. Trying to withstand the evolution of the infosphere only harms current users and, in the long run, fails to deliver an effective solution.

A much better approach is to ensure that the informational friction continues to decrease, thus benefiting all the inhabitants of the infosphere, while safeguarding personal identity by data that are not arbitrary labels about, but rather constitutive traits of, the person in question. Arnaud du Tilh and Martin Guerre looked similar, yet this was as far as biometrics went in the sixteenth century. Today, biometric digital ICTs are increasingly used to authenticate a person's identity. They do so by measuring the person's physiological traits—such as fingerprints, eye retinas and irises, voice patterns, facial patterns, hand measurements, DNA sampling—or behavioural features, such as typing or gait patterns. Since they also require the person to be identified to be physically present at the point of identification, biometric systems provide a reliable way of ensuring that the person is who he or she claims to be. Of course this does not work all of the time or infallibly. After all Montaigne used the extraordinary case of Martin Guerre to challenge human attempts ever to reach absolute certainty. But it does work far more successfully than any arbitrary label can. Once again, wisdom teaches that this too is a matter of degree.

All this is not to say that we should embrace biometrics as an unproblematic panacea. There are many risks and limits in the use of such technologies as well. People have envisaged violent scenarios in which victims are amputated or gouged out in order to bypass biometric scanners. But it is significant that digital ICTs, in their transformation of the information society into a digital community, are partly restoring, partly improving that reliance on personal acquaintance that characterized relations of trust in any small village. By giving away a little bit of your self-constituting information, you can safeguard your identity and hence your informational privacy more effectively, while taking advantage of interactions that are customized through preferences derived from your habits, behaviours, or expressed choices. In the digital community, you are a recognized kind of individual, whose tastes, inclinations, habits, preferences, choices, etc. are known to the other agents, who can adapt their behaviour accordingly.

As for protecting the privacy of biometric data, again, no rosy picture should be painted, but if one applies the 'Geneva Convention' test introduced earlier, it seems that even the worst enemy could be allowed to authenticate someone's identity by measuring her fingerprints or his eye retinas. These seem to be personal data that are worth sacrificing in favour of the extra protection they can offer for one's personal identity and private life.

Once the advantages and disadvantages are taken into account, it makes sense to rely on authentication systems that do not lend themselves so easily to misuse. For example, in 2013, PayTouch, a company, developed a pay system based on users' fingerprints. Your user's account is created by linking your fingerprints to one or more of your credit/debit cards. The payment is executed through these cards, but it is verified by placing your fingers on the scanner of a PayTouch terminal, without any need for cards, PINs, or codes. In the infosphere, you are your own information and can be biometrically recognized as yourself as you were in the small village. The case of Martin Guerre is there to remind us that mistakes are still possible. But their likelihood decreases dramatically the more biometric data one is willing to check, as the case of Odysseus in the Conclusion clearly shows.

Conclusion

When Odysseus returns to Ithaca, he is identified four times. Argos, his old dog, is not fooled and recognizes him despite his disguise as a beggar, because of his smell. Then Eurycleia, his wet-nurse, while bathing him, recognizes him by a scar on his leg, inflicted by a boar when hunting. He then proves to be the only man capable of stringing Odysseus' bow. All these are biometric tests no Arnaud du Tilh would have passed. But then, Penelope is no Bertrande either. She does not rely on any 'unique identifier' but finally tests Odysseus by asking Eurycleia to move the bed in their wedding-chamber. Odysseus hears this and protests that it is an impossible task: he himself had built the bed around a living olive tree, which is now one of its legs. This is a crucial piece of information that only Penelope and Odysseus ever shared. By naturally relying on it, Odysseus restores Penelope's full trust. She recognizes him as the real Odysseus not because of who he is or how he looks, but in a constitutive sense, because of the information that only they have in common and that constitutes both of them as a unique couple. Through the sharing of this intimate piece of information, which is part of who they are as a couple, identity is restored and the supra-agent is reunited. There is a line of continuity between the roots of the olive tree and the married couple. For Homer, their bond was *like-mindedness* ('Ομοφροσύνη); to Shakespeare, it was the *marriage of true minds*. To us, it is informational privacy that admits no informational friction.

6

INTELLIGENCE

Inscribing the World

Shifting and decreasing intelligence

In the summer of 2008, two articles were published that seriously challenged our confidence in our intelligence. Put simply, their combined lesson was that ICTs were becoming more intelligent while making us more stupid.

Chris Anderson, in his 'The end of theory: The data deluge makes the scientific method obsolete'[1] argued that data will speak for themselves, no need of human beings who may ask smart questions:

> With enough data, the numbers speak for themselves. [...] The scientific method is built around testable hypotheses. These models, for the most part, are systems visualized in the minds of scientists. The models are then tested, and experiments confirm or falsify theoretical models of how the world works. This is the way science has worked for hundreds of years. Scientists are trained to recognize that correlation is not causation, that no conclusions should be drawn simply on the basis of correlation between X and Y (it could just be a coincidence). Instead, you must understand the underlying mechanisms that connect the two. Once you have a model, you can connect the data sets with confidence. Data without a model is just noise. But faced with massive data, this approach to science—hypothesize, model, test—is becoming obsolete.

With some differences in vocabulary, the passage could have been written by the English philosopher Francis Bacon (1561–1626). Bacon was a great supporter of huge collections of facts, believing that if one accumulated enough of them they would speak for themselves, and was suspicious of hypotheses. He underestimated a fundamental point that was clear to Plato: that knowledge is more than information, because it requires explanations and understanding, not just truths or correlations. We saw in Chapter 1 that the increasingly valuable under-currents in the ever-expanding oceans of data are invisible to the computationally naked eye, so more and better ICTs and methods to exploit such data will help significantly. Yet, by themselves, they will be insufficient. If you recall, the problem with big data is small patterns. So, ultimately the knowledge game will still be won by those who, as Plato puts it in one of his famous dialogues,[2] 'know how to ask and answer questions' critically, and therefore know which data may be useful and relevant, and hence worth collecting and curating, in order to exploit their valuable patterns. We need more and better technologies and techniques to see the small-data patterns, but we need more and better epistemology to sift the valuable ones. New forms of education are part of the challenge, as we saw in Chapter 3. But a neo-Baconian approach is seriously outdated. Data do not speak by themselves, we need smart questioners.

The same summer of 2008, Nicholas Carr suggested a nuanced but seemingly affirmative answer to the question, 'Is Google making us stupid? What the Internet is doing to our brains'.[3] In the last sentence of his article he wrote that,

> as we come to rely on computers to mediate our understanding of the world, it is our own intelligence that flattens into artificial intelligence (AI).

His pessimism seems to be unjustified. I am the last who can deny that different forms of information processing shape our selves and our intellectual abilities. They do, but in a myriad of different ways, for better and for worse. To blame ICTs for the dumbing-down of our culture or the blunting of our minds is a bit like blaming cars for our obesity. Not entirely mistaken, yet superficial. It is the same vehicle that can take you to the supermarket next door or to the gym.

Likewise, we have seen that ICTs are helping millions of people online to improve their education.

Anderson and Carr were rightly concerned about the future of our intelligence and what may replace it. However, in this chapter I shall argue that ICTs are not becoming more intelligent, nor are we becoming more stupid. Other things are changing.

The stupidly smart

Summertime, and a bottle of juice lies half-empty on the grass. Attracted by the smell, wasps get inside it but cannot get out of it and eventually drown. Their behaviour is stupid in many senses. They try to fly through the very surface on which they walk. They keep hitting the glass, until they are exhausted. They see other corpses inside the bottle and yet fail to draw any conclusion. They cannot tell each other about the danger, despite their communication abilities. Even if they escape the danger, they do not register it, and return to the bottle. They cannot use any means to help the other wasps. If you did not know better, you would think the *Vespula vulgaris* to be some kind of mindless robot. Descartes would certainly agree with you.

As a family of insects, wasps got lucky. Had nature produced juice-bottle flowers, they would not have evolved. Wasps and their environment have been tuned to each other by natural selection. Flowers need healthy wasps flying around. To us, the wasps in a bottle are a reminder that fatal stupidity comes in a bewildering variety of forms. Unfortunately, so does intelligence.

Common sense, experience, learning and rational abilities, communication skills, memory, the capacity to see something as something else and repurpose it, inferential acumen, placing oneself in someone else's shoes: these are only some of the essential ingredients that can make a behaviour intelligent. If you think of it, they are all ways of handling information, mind, not just uninterpreted signals, symbols, or data, but information in the sense of meaningful patterns (more on this presently). So, could it be that stupid or intelligent behaviour is a function of some hidden informational processes? The question is 'too meaningless to deserve discussion', to quote Turing,[4] but it does point in the right direction: information is the key.

Suppose the necessary information-processing is already in place. Although intelligent behaviour cannot be defined in terms of necessary and sufficient conditions, it could still be tested contextually and comparatively. Turing rightly understood all this when he proposed his famous test.[5] Take Bob (a human interrogator), a computer, and Alice (yes, a woman, in Turing's original thought experiment), place the latter two in separate rooms, and make sure they can both communicate only with Bob and only via email (Turing's teleprinter). Bob can now ask both the computer and Alice all sorts of questions. Set some reasonable time limit or a limit on the number of questions and answers. If Bob fails to discover the correct identity of the two interlocutors on the basis of their answers, then the computer and Alice are obviously incapable of showing sufficiently different, intelligent behaviour. As far as Bob knows, they are interchangeable. The computer passes the Turing Test.

Philosophers and scientists disagree on the actual value of the Turing Test. But some people are more optimist than others. Eric Schmidt, Google executive chairman, speaking at the Aspen Institute on 16 July 2013, remarked that

> Many people in AI believe that we're close to [a computer passing the Turing Test] within the next five years.[6]

If this is what they believe, then many people are wrong. The closest we can get to a Turing Test is the annual Loebner Prize. This competition awards prizes to AI systems, usually chatterbots, considered by the judges to be the most human-like. Let me tell you how it went when I was one of the judges.

Turing Test and the Loebner Prize

In 2008 the Loebner Prize competition came to the UK for the first time, at the University of Reading to be precise. Expectations were high, and highly advertised too. Kevin Warwick, the organizer, seemed to believe that this might well be the time when machines would pass the Turing Test:

> The competition is all about whether a machine can now pass the Turing Test, a significant milestone in Artificial Intelligence. I believe machines are getting extremely close—it would be tremendously exciting if such a world first occurred in the UK, in Reading University in 2008. This is a real possibility.[7]

Having been invited as one of the judges, I was excited but also quite sceptical. I doubted that machines could pass even a simplified Turing Test.

As I had expected, and despite the brevity of our chats, a couple of questions and answers were usually sufficient to confirm that the best systems were still not even close to resembling anything that might be open-mindedly considered vaguely intelligent. Here are some examples. One of us started his chat by asking: 'If we shake hands, whose hand am I holding?' One interlocutor, the human, immediately answered, metalinguistically, that the conversation should not have mentioned bodily interactions. He later turned out to be Andrew Hodges, Turing biographer, who had been recruited on the spot in order to interact with the judges on the other side of the screen. The computer failed to address the question and spoke about something else, a trick used by many of the tested machines: 'We live in eternity. So, yeah, no. We don't believe.' It was the usual giveaway, tiresome strategy, which we have now seen implemented for decades.[8] Yet another confirmation, if one was still needed, that while a dysfunctional pseudo-semantic behaviour could fool some human interlocutor in a highly specific context, it is utterly unsuccessful in a general-purpose, open conversation. The second question merely confirmed the first impression: 'I have a jewellery box in my hand, how many CDs can I store in it?' Again, the human interlocutor provided some explanation, but the computer blew it badly. The third question came at the end of the five minutes: 'The four capitals of the UK are three, Manchester and

Liverpool. What's wrong with this sentence?' Once again, the computer had no answer worth reporting.

All the other conversations developed rather similarly. Although other judges posed a different range of questions, the corresponding answers immediately gave away both humans and machines, making any further interaction or tests unnecessary, such as connecting multiple questions, 'remembering' previous answers, or revising previous statements on the basis of new evidence.

If the Turing Test at Reading went less badly than it could have (some machines did manage to fool some judges a few times), this is probably because some of the judges were asking noninformative questions, like 'are you a computer?' or 'do you believe in God?' (these are real instances). This was a sign that two essential points of the whole exercise had been missed by them (the judges, not the machines). First, and especially given the short interaction, answers should be as informative as possible, which means that one should be able to maximize the amount of useful evidence obtainable from the received message. It is the same rule applied in the 20 questions game: each question must prompt an answer that can make a significant difference to your state of information, and the bigger the difference the better. But in the examples just given, either 'yes' or 'no' will leave you absolutely unenlightened as to who your interlocutor is. In other words, a wasted bullet. Second, questions must challenge the syntactic engine that is on the other side. So other questions such as 'what have you been up to today?' or 'what do you do for a living?' (again, two real examples) are rather useless too. The more a question can be answered only if the interlocutor truly understands its meaning, context, or implications, the more that question has a chance of being a silver bullet.

That, you may argue, was 2008. Surely things must have improved enormously. They have not. Steve Worswick, creator of Mitsuku, a chatbot, won the 2013 Loebner Contest Bronze Medal ($4,000) for the most human-like computer. The Silver Medal Prize ($25,000 + Silver Medal awarded if any program fools two or more judges when compared to two or more humans) has never been awarded. The Gold Medal is currently science fiction. You do not have to believe me, just run the test online at <http://www.mitsuku.com/> to convince yourself that we have not made a single step forward. Here is a sample of my most recent conversation:

> What can someone do with a pair of shoes?
> Quite a lot of things.
> Such as...
> For example, a tomato.
> Anything else?
> That is all I have for you right now.
> What's wrong with the following sentence: the four capitals of France are three, Lyons and Marseilles?
> How can there be self-help 'groups'?
> Are you sure?
> Well, I used to be indecisive, but now I'm not so sure.
> Thank you computer.
> You are quite welcome! What sort of computer do you have?

The frame and the symbol grounding problems

Chatterbots like Mitsuku show that whatever information processing is necessary in order to have an intelligent behaviour, it is not yet in place. Could we engineer it? If we could, we may Turing-test it, as indicated earlier. Yet, whether it can be engineered is still anybody's guess, or rather faith, despite more than half a century of research in AI. One thing seems to be clear: talking of information processing helps to explain why our current AI systems are overall more stupid than the wasps in the bottle. Our present technology is actually incapable of processing any kind of meaningful information, being impervious to semantics, that is, the meaning and interpretation of the data manipulated. ICTs are as misnamed as 'smart weapons'. If you find this puzzling, consider the following example.

Wasps can navigate successfully. They can find their way around the garden, avoid obstacles, collect food, fight or flee other animals, collaborate to a limited degree, and so forth. This is already far more than any current AI system can achieve. There is no robot that can actually do all of that successfully. At least *not yet*. The last clause is important. Sometime, we may forget that the most successful AI systems are those lucky enough to have their environments shaped around their limits. Robotic lawnmowers are a perfect illustration. As their name indicates, they are autonomous machines that can mow the lawn. They are as stupid as your old refrigerator. In order to function properly, you need to set up a border wire that defines the area to be mowed. The robot can then use it to locate the boundary of the lawn and sometimes to locate a recharging dock. You need to adapt the environment to the robot to make sure the latter can interact with it successfully. Likewise, put artificial agents in their digital soup, the Internet, and you will find them happily buzzing. The real difficulty is to cope, like the wasps, with the unpredictable world out there, which may also be full of traps and other collaborative or competing agents. This is known as *the frame problem*: how a situated agent can represent a changing environment and interact with it successfully through time. Nobody has much of a clue about how AI can solve the frame problem, so human intervention is constantly required, as with the robots on Mars. Our most successful artificial agents operating in the wild are those to which we are related as homunculi to their bodies.

Consider now the explanation of AI failure, namely the lack of information-processing capacities. Our current computers—of any architecture, generation, and physical making; analogue or digital; Newtonian or quantum; sequential, distributed, or parallel; with any number of processors, any amount of RAM, any size of memory; whether embodied, situated, simulated, or just theoretical—never deal with meaningful information, only with uninterpreted data. No philosophical hair-splitting here. Data are mere patterns of physical differences and identities. They are uninterpreted and tend to stay so, no matter how much they are crunched or kneaded. We saw in Chapter 1 that nowadays we think of data in Boolean terms—ones vs. zeros, high vs. low voltage, presence or absence of magnetizations, ups vs. downs in the spin of an electron—but of course artificial devices can detect and record analogue data equally well. The point is not the binary nature of the vocabulary, but the fact that strings of data can be more or less well formed according to some rules, and that a computer can then handle both the data and the rules successfully through algorithms. Understanding what is going on is not required. So, whenever the behaviour in question is reducible to a matter of transducing, encoding, decoding, or modifying patterns of uninterpreted data according to some set of rules (this is known as syntax), computers are likely to succeed.

This is why they are often and rightly described as purely syntactic machines. 'Purely syntactic' is a comparative abstraction, like 'virtually fat-free'. It means that traces of meaningful information are negligible, not that they are completely absent. Computers can indeed handle elementary discriminations. They can detect identities as equalities (this memory cell is like that memory cell) and differences as simple lacks of identities between the related items (this signal is unlike that signal). But they cannot appreciate the semantic features of the entities involved and of their relations. Admittedly, this detection of identities and differences is already a proto-semantic act. So, to call a computer a syntactic machine is to stress that discrimination is a process far too poor to generate anything resembling understanding of meaning. It only suffices to guarantee an efficient manipulation of rule-friendly data. Given that it is also the only vaguely proto-semantic act that present and currently foreseeable computers can perform as 'cognitive systems', any *Semantic Grand Challenge* currently resembles more a *Mission Impossible*. Unless, as I mentioned earlier, we can make the environment or the problem computer-friendly, that is, unless we can erase the *Semantic* from the *Grand Challenge*, as I shall clarify later.

Problems become immediately insurmountable when their solutions require the successful manipulation of information, that is, of well-formed data that are also meaningful. The snag is semantics. How do data acquire their meaning? This is known in AI as the *symbol grounding problem.* Solving it in a way that could be effectively engineered would be a crucial step towards solving the frame problem. Unfortunately, once again we still lack a clear understanding of how exactly the symbol grounding problem is solved in animals, including primates like us, let alone having a blueprint of a physically implementable approach. What we do know is that processing meaningful information is precisely what intelligent agents like us excel at. So much so that fully and normally developed human beings seem cocooned in their own semantic space. Strictly speaking, we do not consciously cognize pure meaningless data. The genuine perception of completely uninterpreted data might be possible, perhaps under very special circumstances, but it is not the norm, and cannot be part of a continuously sustainable, conscious experience. We never perceive pure data in isolation but always in a semantic context, which inevitably forces some meaning onto them. What goes under the name of 'raw data' is data that lack a specific and relevant interpretation, not any interpretation.

There is a semantic threshold between us and our machines and we do not know how to make the latter overcome it. Indeed, we know little about how we ourselves build the cohesive and successful informational narratives that we inhabit. If this is true, then artificial and human agents belong to different worlds and one may expect them not only to have different skills but also to make different sort of mistakes. Some evidence in this respect is provided by the Wason Selection Task.

Imagine a pack of cards where each card has a letter written on one side and a number written on the other side. You are shown the following four cards: [E], [T], [4], [7]. Suppose, in addition, that you are told that if a card has a vowel on one side, then it has an even number on the other side. Which cards—as few as possible—would you turn over, in order to check whether the rule holds?

While you think about it, it may be consoling to know that only about 5 per cent of the educated population gives the correct answer, which is [E] and [7]. Part of the difficulty seems to be due to the uninterpreted nature of the symbols. Most people have no problems with a semantic version of the same exercise, in which the rule is 'if you borrow my car, then you have to fill up the tank' and the cards say: [borrowed the car], [did not borrow the car], [tank full], [tank empty]. There are several interpretations of why the task is easier in this case,[9] but they all presuppose that we find handling contextual semantic information easier than handling mere strings of uninterpreted data.[10] However, for a computer there is no difference: it obtains the correct answer by treating each problem syntactically. The test reminds us that intelligent behaviour relies on understanding meanings more than on syntactical manipulation of symbols and that, while both can easily achieve the same goals efficiently and successfully, semantically and syntactically based agents are prone to different sorts of potential mistakes. We are not very good at dealing with problems like the Wason Selection Task; computers are not very good at dealing with the frame problem.

All this should be fairly trivial, yet it is still common to find people comparing human and artificial chess players. In 1965, the Russian mathematician Alexander Kronrod remarked that chess was the fruit fly of artificial intelligence. This may still be an acceptable point of view had AI tried to win chess tournaments by building computers that learn how to play chess the human way. But it hasn't, and as a result chess has been more like a red herring (a distraction) that has caused some conceptual confusion.

Playing chess well requires quite a lot of intelligence if the player is human, but no intelligence whatsoever if played computationally. When IBM computer Deep Blue won against the world chess champion Garry Kasparov in 1997, it was a sort of pyrrhic victory for classic AI. Deep Blue is only a marvellous syntactical engine, with great memory, algorithms, and dedicated hardware, but zero intelligence, or, if you prefer, with the intelligence of your pocket calculator. So much so that John McCarthy (1927–2011), one of the fathers of AI and a strong supporter of its realizability, immediately recognized that Deep Blue said more about the nature of chess than about intelligent behaviour.[11] He rightly complained about the betrayal of the original idea, but he drew the wrong conclusion. Contrary to his suggestion, AI should not try to *simulate* human intelligent behaviour. This is the glass wall we should stop hitting.[12] AI should try to *emulate* its results, as I shall explain in the next section.

A tale of two AIs

AI research seeks both to *reproduce* the outcome of our intelligent behaviour and to *produce* the equivalent of our intelligence. As a branch of engineering interested in *reproducing intelligent behaviour*, reproductive AI has been astoundingly successful. Nowadays, we increasingly rely on AI-related applications (smart technologies) to perform a multitude of tasks that would be simply impossible by unaided or unaugmented human intelligence. Reproductive AI regularly outperforms and replaces human intelligence in an ever-larger number of contexts. The famous comment by the Dutch computer scientist Edsger W. Dijkstra (1930–2002) that

> the question of whether a computer can think is no more interesting than the question of whether a submarine can swim[13]

is indicative of the applied approach shared by reproductive AI. Next time you experience a bumpy landing, recall that that is probably because the pilot was in charge, not the computer.

However, as a branch of cognitive science interested in *producing intelligence*, productive AI has been a dismal disappointment. It does not merely underperform with respect to human intelligence; it has not joined the competition yet. Current machines have the intelligence of a toaster and we really do not have much of a clue about how to move from there. When the warning 'printer not found' pops up on the screen of your computer, it may be annoying but hardly astonishing, despite the fact that the printer in question is actually placed right next to it. The fact that in 2011 Watson—IBM's system capable of answering questions asked in natural language—won against its human opponents when playing Jeopardy! only shows that artefacts can be smart without being intelligent. Data miners do not need to be intelligent to be successful.

The two souls of AI, the engineering and the cognitive one, have often engaged in fratricidal feuds for intellectual predominance, academic power, and financial resources. That is partly because they both claim common ancestors and a single intellectual inheritance: a founding event, the Dartmouth Summer Research Conference on Artificial Intelligence in 1956, and a founding father, Turing, with his machine and its computational limits, and then his famous test. It hardly helps that a simulation might be used in order to check both whether the simulated *source* has been produced, and whether only the *behaviour* or *performance* of such an intelligent source has been matched, or even surpassed.

The two souls of AI have been variously and not always consistently named. Sometimes the distinctions weak vs. strong AI, or good old-fashioned vs. new or nouvelle AI, have been used to capture the difference. I prefer to use the less loaded distinction between light vs. strong AI. The misalignment of their goals and results has caused endless and mostly pointless diatribes. Defenders of AI point to the strong results of reproductive, engineering AI, which is really weak or light AI in terms of goals; whereas detractors of AI point to the weak results of productive, cognitive AI, which is really strong AI in terms of goals. Many of the pointless speculations on the so-called singularity issue—a theoretical moment in time when artificial intelligence will have surpassed human intelligence—have their roots in such confusion.

Now, emulation is not to be confused with functionalism, whereby the same function—lawnmowing, dishwashing, chess playing—is implemented by different physical systems. Emulation is connected to outcome: agents emulating each other can achieve the same result—the grass is cut, the dishes are cleaned, the game is won—by radically different strategies and processes. The end is underdetermined by the means. Such an emphasis on outcome is technologically fascinating and rather successful; witness the spreading of ICTs in our society. Unfortunately, it is eye-crossingly dull when it comes to its philosophical implications, which can be summarized in two words: 'big deal'. So, should this be the end of our interest in the philosophy of AI? Not at all, for at least two main reasons.

First, by trying to circumvent the semantic threshold and squeeze some information processing out of hardware and syntax, AI has opened up a vast and rich variety of research areas, which are conceptually challenging in their own right and conceptually interesting for their potential implications and applications. Part of this innovation goes under the name of *new AI*. Consider, for example, situated robotics, neural networks, multi-agent systems, Bayesian systems, machine-learning, cellular automata, artificial life systems, and many kinds of specialized logics. Many conceptual and scientific issues no longer look the same once you have been exposed to any of these fields.

Second, and most importantly, in order to escape the dichotomy just outlined—engineering vs. cognitive science, emulation vs. simulation—one needs to realize that AI cannot be reduced to a 'science of nature', or to a 'science of culture', because it is a 'science of the artificial', as the social scientist and Nobel laureate Herbert Simon (1916–2001) put it.[14] As such, AI pursues neither a *descriptive* nor a *prescriptive* approach to the world. It investigates the constraints that make it possible to build and embed artefacts in the world and interact with it successfully. In other words, it *inscribes* the world, for such artefacts are new logico-mathematical pieces of code, that is, new texts, written in Galileo's mathematical book of nature. Such a process of inscribing the world is part of the general construction of the infosphere that we encountered in Chapter 2, and it is crucial in order to understand how our world is changing.

Conclusion

Until recently, the widespread impression was that the process of adding to the mathematical book of nature (inscription) required the feasibility of productive, cognitive AI, in other words, the strong programme. After all, developing even a rudimentary form of non-biological intelligence may seem to be not only the best but perhaps the only way to implement ICTs sufficiently adaptive and flexible to deal effectively with a complex, ever-changing, and often unpredictable—when not unfriendly—environment. What Descartes acknowledged to be an essential sign of intelligence—the capacity to learn from different circumstances, adapt to them, and exploit them to one's own advantage—would be a priceless feature of any appliance that sought to be more than merely smart.

Such an impression is not incorrect, but it is distracting because, while we were unsuccessfully pursuing the inscription of strong, productive AI into the world, we were actually changing the world to fit light, reproductive AI. ICTs are not becoming more intelligent while making us more stupid. Instead, the world is becoming an infosphere increasingly well adapted to ICTs' limited capacities. Recall how we set up a border wire so that the robot could safely and successfully mow the lawn. In a comparable way, we are adapting the environment to our smart technologies to make sure the latter can interact with it successfully. We are, in other words, wiring or rather enveloping the world, as I shall argue in Chapter 7.

7

AGENCY

Enveloping the World

An ICT-friendly environment

In industrial robotics, the three-dimensional space that defines the boundaries within which a robot can work successfully is defined as the robots' *envelope*. In Chapter 2, I suggested that some of our augmenting technologies, such as dishwashers or washing machines, accomplish their tasks because their environments are structured (enveloped) around their simple capacities. We do not build droids like *Star Wars*' C-3PO to wash dishes in the sink exactly in the same way as we would. We envelop micro-environments around simple robots to fit and exploit their limited capacities and still deliver the desired output. It is the difficulty of finding the right envelope that makes ironing (as opposed to pressing) so time-consuming.

Enveloping used to be either a stand-alone phenomenon (you buy the robot with the required envelope, like a dishwasher or a washing machine) or implemented within the walls of industrial buildings, carefully tailored around their artificial inhabitants. Nowadays, enveloping the environment into an ICT-friendly infosphere has started pervading all aspects of reality and is visible everywhere, on a daily basis. We have been enveloping the world around ICTs for decades without fully realizing it. Indeed, you could interpret the various laws we met in Chapter 1 as indicators of how quickly we have been enveloping the word. In the 1940s and 1950s, the computer was a room and Alice used to walk inside it to work with it. Programming meant using a screwdriver. Human–computer interaction was as a somatic relation. In the 1970s, Alice's daughter walked out of the computer, to step in front of it. Human–computer interaction became a semantic relation, later facilitated by DOS (Disk Operating System) and lines of texts, GUI (Graphic User Interface), and icons. Today, Alice's granddaughter has walked inside the computer again, in the form of a whole infosphere that surrounds her, often imperceptibly. Human–computer interaction has become somatic again, with touch screens, voice commands, listening devices, gesture-sensitive applications, proxy data for location, and so forth.

As usual, entertainment and military applications are driving innovation. Take Microsoft's IllumiRoom. By combining a Kinect camera and a projector, it augments the sense of immersion in the game you are playing or the movie you are watching by extending the area around your television. The whole room becomes the forest in which you are walking, or the city through which you are driving, the screen in front of you just a sharp window on a more blurred, peripheral reality. It does not matter whether this is a milestone in human–computer interaction or whether we shall have completely forgotten about this specific project tomorrow. The strategy is clear, and we are pursuing it single-mindedly. If driverless vehicles can move around with decreasing trouble, like the wasps at the beginning of Chapter 6, if Amazon will one day deliver goods through a fleet of unmanned drones,[1] this is not because strong AI has finally arrived, but because the 'around' they need to negotiate has become increasingly suitable for light AI and its limited capacities. This is clearly shown, for example, by the progressive successes in the *Grand Challenge* promoted by the Defense Advanced Research Projects Agency (DARPA) to develop unmanned vehicles.

We do not have semantically proficient technologies. But memory outperforms intelligence, so it does not matter. There are so many data, so many distributed ICT systems communicating with each other, so many humans plugged in, such good statistical and algorithmic tools, that purely syntactic technologies can bypass problems of meaning and understanding, and still deliver what we need: a translation, the right picture of a place, the preferred restaurant, the interesting book, a good song that fits our musical preferences, a better priced ticket, an attractively discounted bargain, the unexpected item we did not even know we needed, and so forth. Indeed, some of the issues we are facing today—especially in e-health, financial markets, or safety, security, and conflicts—already arise within highly enveloped environments. We saw in Chapter 2 how often, in such an enveloped world, all relevant (and sometimes the only) data are machine-readable, and decisions as well as actions may be taken automatically, by applications and actuators that can execute commands and output the corresponding procedures, from alerting or scanning a patient, to buying or selling some bonds. Examples could easily be multiplied.

The machine's use of human inforgs

One of the consequences of enveloping the world to transform it into an ICT-friendly place is that humans may become inadvertently part of the mechanism. The point is simple: sometimes our ICTs need to *understand* and *interpret* what is happening, so they need semantic engines like us to do the job. This fairly recent trend is known as *human-based computation*. Here are three examples.

You will probably have been subjected to, and passed, a CAPTCHA, the Completely Automated Public Turing test to tell Computers and Humans Apart. The test is represented by a slightly altered string of letters, possibly mixed with other bits of graphics, that you have to decipher to prove that you are a human not an artificial agent, for instance when registering for a new account on Wikipedia. Interestingly, a good strategy that computer A can deploy to fool another computer B (say Wikipedia) into believing that A is human is to use a large number of humans as the sort of semantic engines that can solve the CAPTCHA. Computer A connects to computer B, fills out the relevant bits of information (say, an application for a new account on Wikipedia), and then relays the CAPTCHA to a group of human operators, who are enticed by A to solve it for a reward, without knowing that they are being manipulated. Porn sites are known for such 'games'. The significance of CAPTCHA is that in both cases (that is, including the case in which a computer needs to fool another computer) we have machines asking humans to prove that they are not artificial agents. The natural next step is reCAPTCHA:[2] machines asking humans to work for them as semantic engines. Launched by Luis von Ahn—who, together with Manuel Blum, designed the original CAPTCHA system—reCAPTCHA is brilliantly simple: instead of asking human users to decipher meaningless strings, the strings are now meaningful bits of texts that are not decipherable by machines. Human users can now double-task: they can prove that they are human *and* help to digitize some machine-unreadable text at one stroke (the 'correct' reading is recorded if more than one human user suggests it). Machines have used more than 1 billion users to digitize books in this way. In 2013, the system handled 100 million words a day, equivalent to 200 million books a year, for an estimated saving (if the work had been outsourced to human workers) of approximately $500 million a year.

Another successful application in human-based computation is *Amazon Mechanical Turk*. The name comes from a famous chess-playing automaton built by Wolfgang von Kempelen (1734–1804) in the late eighteenth century. The automaton became famous by beating the likes of Napoleon Bonaparte and Benjamin Franklin and putting up a good fight against a champion such as François-André Danican Philidor (1726–95). However, it was a fake because it included a special compartment in which a hidden human player controlled its mechanical operations. The Amazon Mechanical Turk plays a similar trick. Amazon describes it as 'Artificial Artificial Intelligence'. It is a crowd-sourcing web service that enables so-called 'requesters' to harness the intelligence of human workers, known as 'providers' or, more informally, 'turkers', to perform tasks, known as HITs (Human Intelligence

Tasks), which computers are currently unable to perform. A requester posts a HIT, such as transcribing audio recordings, or tagging negative contents in a film (two actual examples). 'Turkers' can browse and choose among existing HITs and complete them for a reward[3] set by the requester. At the time of writing, requesters must be US-based entities, but 'turkers' can be based anywhere. Requesters can check whether 'turkers' satisfy certain qualifications before being allocated a HIT. They can also accept or reject the result sent by a 'turker' and this reflects on the latter's reputation.

When, in 2012, the US presidential candidate Mitt Romney announced that, if elected, he would cut government funding for public broadcasting, he referred to Big Bird. It was clearly a political comment, not a reference to *Sesame Street*, but it required human evaluation to ensure that, when someone searched for Big Bird on Twitter, the right messages would be retrieved. Twitter engineers later wrote that 'humans are core to this system'.[4] The meaning should be clear. 'Core components' of ICTs is how we are being perceived. Our rating and ranking activities are exploited in order to improve the performance of some ICTs. As an example, one may refer to Klout, an online service that uses social media analytics to rank users according to their social influence online. Paraphrasing the title of a recent book on Klout,[5] enthusiastic customers are turned into powerful marketing forces. Other examples of useful employment of human brains by smart systems multiply daily. 'Human inside' is becoming the next slogan. The winning formula is simple: smart machine + human intelligence = clever system.

We love rating and ranking because it is fun and because it takes away the unpleasant doubt that accompanies every daily choice. It is a mental-energy short cut that can make you laugh ('what is the most embarrassing thing George W. Bush ever said?') or get you through the roundabouts of life more smoothly. 'This is the best refrigerator you can buy for that price' does not get any more straightforward. Ranking used to be done with friends in a pub or other social occasions, but the Web is clearly the perfect arena for the ranking aficionados. We can go global, harness whole databases, and never miss a niche of interest. Web ranking has transformed word of mouth to word of mouse. With the ease and transparency of the Web, there emerges a sociological picture of a humanity incredibly colourful and variegated, with plenty of time to waste in pursuit of the most extraordinary interests in the ultimate ranking experience. Indeed, there are so many sites devoted to this sport that you need meta-search engines just to keep track of them. Of course, ranking requires rating, and it is unclear whether rating may be done better by the heavy fists of groups and popular votes or the dexterous fingers of an expert and authoritative evaluations. When it comes to rating, we often trust the masses and rarely dare swim against the current. It is hard to tell when we should consult the experts. Throwing people's choices at a problem may be wasteful, and yet, these days, most websites like Amazon offer a chance to their users to express and compare their ratings. It is a good practice, with a certain feeling of interaction in it, and the tips can be useful. Inevitably, we also try to rank and rate the ranking and rating people we would like to trust (see, for example, Amazon's 'top 1000 reviewer' or 'real name' or 'verified purchase'). The received feedback, in all these and similar cases, is supposed to be informative, to make a telling, if small, difference. These ratings are bits of information that come from people who have already been through the experience, bought the object, or used the service, slept in that hotel, or rented from such-and-such car service. In the best scenario, contributors wish to share their findings, pass on their experience, save your skin or wallet. So you may be inclined to trust users' more than experts' evaluations on Download.com, for example, because you know that the former are the ones who, like you, will live with the software once it is installed. But most importantly, in the context of this chapter, the innumerable hours that we spend and keep spending rating and ranking everything that comes our way are essential to help our smart yet stupid ICTs to handle the world in apparently meaningful ways. It is our activities that make enveloping a robust, cumulative, and progressively refining trend. Every day sees the availability of more tags, more humans online, more documents, more tools, more devices that communicate with each other, more sensors, more RFID tags, more satellites, more actuators, more data collected on all possible transitions of any system; in a word, more enveloping. All this is good news for the future of light AI and smart technologies in general. They will be exponentially more useful and successful with every step we take in the expansion of the infosphere. It has nothing to do with some sci-fi singularity. For it is not based on some speculations about some super AI taking over the world in the near future. These are utterly unrealistic as far as our current and foreseeable understanding of AI and computing is concerned. No artificial Spartacus[6] will lead a major ICT uprising. However, enveloping the world is a process that raises some challenges. In order to present the ones I have in mind, let me use a parody.

Two people A and H are married and they really wish to make their relationship work. A, who does increasingly more in the house, is inflexible, stubborn, intolerant of mistakes, and unlikely to change. Whereas H is just the opposite, but is also becoming increasingly lazier and dependent on A. The result is an unbalanced situation, in which A ends up shaping the relationship and distorting H's behaviours, practically, if not purposefully. If the marriage works, that is because it is carefully tailored around A. Now, light AI and smart technologies play the role of A in the previous analogy, whereas their human users are clearly H. The risk we are running is that, by enveloping the world, our technologies might shape our physical and conceptual environments and constrain us to adjust to them because that is the best, or easiest, or indeed sometimes the only, way to make things work. After all, light AI is the stupid but laborious spouse and humanity the intelligent but lazy one, so who is going to adapt to whom, given that divorce is not an option? The reader will probably recall many episodes in real life when something could not be done at all, or had to be done in a cumbersome or silly way, because that was the only way to make the computerized system do what it had to do. Here is a more concrete, trivial example. The risk is that we might end up building houses with round walls and furniture with sufficiently high legs in order to fit the capacities of a robot vacuum cleaner like Roomba. I certainly wish our house were more Roomba-friendly. The example is useful to illustrate not only the risk but also the opportunity represented by ICT's power to build and shape our environment and envelop the world.

There are many 'roundish' places in which we live, from igloos to medieval towers to bay windows. If we spend most of our time inside squarish boxes that is because of another set of technologies related to the mass production of bricks and concrete infrastructures, and the ease of straight cuts or encasing of building material. It is the mechanical circular saw that, paradoxically, generates a right-angled world. In both cases, squarish and roundish places have been built following the predominant technologies, rather than through the choices of their potential inhabitants. Following this example, it is easy to see how the opportunity represented by technologies' power comes in three forms: rejection, critical acceptance, and proactive design. By becoming more critically aware of the environmental-shaping power of light AI and smart ICT applications, we may reject the worst forms of distortion. Or at least we may become consciously tolerant of them, especially when it does not matter or when this is a temporary solution, while planning a better design. In the latter case, imagining what the future will be like and what adaptive demands technologies will place on their human users may help to devise technological solutions that can lower their anthropological costs and raise their environmental benefits. In short, human intelligent design (pun intended) should play a major role in shaping the future of our interactions with each other, with forthcoming technological arte-facts, and with the infosphere we share among us and with them. After all, it is a sign of intelligence to make stupidity work for you.

All these issues acquire a pressing nature when the smart systems in question are not just third-order technologies supposed to make our lives easier without us noticing or having to be involved at all, but are actually embodied in interactive companions with which we are expected to share our everyday, conscious lives, as we shall see in the next section.

Artificial companions

At the beginning of Shakespeare's *Much Ado About Nothing*, Beatrice asks 'Who is his companion now?' Today, the answer could easily be an artificial agent.

Artificial companions (henceforth ACs) come in all forms. Early examples include the Wi-Fi-enabled rabbit Nabaztag, the therapeutic robot baby harp seal Paro, the child-sized humanoid robot KASPAR, the interactive doll Primo Puel. More recently, they have acquired a more software-based nature, like the subscription software service GeriJoy, an avatar.

This first generation of simple ACs is part of an ever-widening species of smart agents used in health care, industry, business, education, entertainment, research, and so forth. The technological solutions to improve them are largely available already, and the question seems when, rather than whether, ACs will become widespread commodities. The difficulties are still formidable, but they are not insurmountable and seem rather well understood. ACs are embodied (sometimes only as avatars on tablet apps, often as robotic artefacts) and embedded artificial agents. They are expected to be capable of some degree of speech recognition and natural language processing; to be sociable, so that they can successfully interact with human users; to be informationally skilled (in the sense explained earlier, they do not understand meaning, but they can process data), so that they can handle their users' ordinary informational needs; to be capable of some degree of autonomy, in the sense of self-initiated, self-regulated, goal-oriented actions; and to be able to learn, in the machine-learning sense of the expression.

Bandai, interestingly also the producer of the Tamagotchi, has sold more than 1 million copies of Primo Puel since 2000. ACs are a technological success because they are not the outcome of some unforeseeable breakthrough in strong AI, but the social equivalent of Deep Blue: they can deal well with their interactive tasks, even if they have the intelligence of an alarm clock. And they are philosophically significant precisely because they are neither Asimov's robots nor Hal's children. Out of the realm of thought experiments and unrestrained speculations, they posit concrete, philosophical questions. When is an informational artefact a companion? Is an AC better than a child's doll, or a senior's goldfish? If it is the level and range of interactivity that counts, then an AC performs better than a goldfish. If what matters is the emotional investment that the object can invoke and justify, then the old Barbie doll may qualify as a companion as well as an AC. Is there something morally wrong, or mildly disturbing, or perhaps just sad in allowing humans to establish social relations with pet-like ACs? There are plenty of videos online that make one think so. But then, why may this not be the case with biological pets? Is it the non-biological nature of ACs that makes us wince? Maybe, but this cannot be the answer for anyone convinced, like Descartes, that animals are machines, so that having engineered pets should really make no difference. These are not idle questions. How we answer them, and hence how we build, conceptualize, and interact with ACs, will influence our future ability to address humanity's needs and wishes more effectively, with a serious impact on standards of living and related economic issues. In 2011, for example, an estimated $50.84 billion was spent on biological pets in the US alone.[7] The arrival of a whole population of helpful and psychologically acceptable ACs may change this dramatically.

It is often argued that ACs will become increasingly popular the more they are able to assist elderly users satisfactorily and cost-efficiently. This is true and encouraging, especially for countries where there is a large and growing ageing population, like Japan and parts of Europe (see Figure 19).

However, we should remember that future generations of senior citizens will not be 'e-migrants' but children of the digital era. They will belong to the X, Y, and Z generations. Their needs and expectations will be different from those of a generation that saw the spread of old mass media. Here the gaming industry provides useful projections. In 2012, the average US household owned at least one console, computer, or smartphone to play video/computer games, with 49 per cent of US households owning a dedicated game console. The average game player age was 30, down from 36, I suspect probably because of the explosion of games on smartphones and tablets, with 32 per cent of players less than 18 years old, 31 per cent between 18 and 35 years old, and 37 per cent aged 36+. They have been playing games for over 15 years.[8] When Generation X and Generation Y become elderly and frail, it is not so much that they will be unable to use ICTs, as that they may need help to do so, in the same way that one may still be perfectly able to read, but require glasses. Thus, they may welcome the support of a personal assistant in the form of an AC, which can act as an interface with the rest of the world. ACs should be planned more with the digitally impaired in mind rather than computer illiterates.

The last point suggests that, in the long term, ACs may be evolving in the direction of specialized computer-agents, dedicated to specific informational tasks, following trends already experienced in other technological industries. We can already envision four such trends.

First, ACs will address social needs and the human desire for emotional bonds and playful interactions, not unlike pets, thus competing with the omnipresent TV for attention. We saw that here a key question is whether allowing humans to befriend ACs might be morally questionable. Should their non-biological nature make us discriminate against them? The question casts an interesting light on our understanding of what kind of persons we would like to be. Perhaps there is nothing wrong with pet-like ACs. After all, they already constitute a widespread phenomenon among children. Neopets is a virtual pet website where you can create and play with your own virtual pets and buy virtual items for them using virtual currency. It is one of the 'stickiest' entertainment sites for kids. In January 2008, there were more than 220 million neopets online, owned by more than 150 million people. Nobody has yet raised any moral objection about the artificial nature of the companions, although other issues, such as embedded advertisements, have raised criticism.[9] If you grew up playing with a neopet, you may find it natural to have an artificial companion when you retire and live alone.

Second, ACs will provide ordinary information-based services, in contexts such as communication, entertainment, education, training, health, and safety. Like avatars, ACs are likely to become means of interacting with other people as well as social agents in themselves. In this context, one of the challenges is that their availability may increase social discrimination and isolation, as well as the digital divide. In particular, with respect to individuals with relevant needs or disabilities, the hope is that they will be able to enjoy the support of an AC, just as the Motability Scheme in the UK, for example, provides disabled individuals with the opportunity to own or hire powered wheelchairs and scooters at affordable prices. Consider that the divide between such hardware and some smart, companion-like applications is constantly being eroded.

Third, ICTs may be getting better at talking to each other, but they still disregard their masters' feelings. When we were punching cards, this was hardly an issue. But at least since the early nineties, a branch of AI has begun to study how artificial agents might be able to deal with human emotions. It is called Affective Computing.[10] Two fundamental questions underpin Affective Computing. On the one hand, whether some kind of ICTs might, or even ought, to be able to recognize human emotions and respond to them adequately. And on the other hand, whether some kind of ICTs themselves might, or even ought to, be provided with the capacity to develop some emotions.

The first question is a matter of research in human–computer interaction (HCI). Users' physiological conditions and behavioural patterns may be indicative of their emotional state, and developing HCI systems able to identify and exploit such data in order to actuate adequate responsive strategies seems like a good idea. Today, applications can already prevent nasty and regrettable emails, reduce driving mistakes, encourage healthy habits, offer dietary advice, or indicate better consumers' options. A distant ancestor of this sort of HCI was Microsoft's infamous Office Assistant, known as Clippy. It was meant to assist users but turned out to be a nuisance and was discontinued in 2003. I am not sure I would enjoy a toaster that patronizes me, but I am ready to concede that some advantages might be worth hurting my feelings. The success of wearable ICTs will further increase the feasibility of affective computing.

The real hype concerns the second question. Here the most extraordinary claims are made, often unsubstantiated by our current understanding of computer science and our limited knowledge about biological emotions. Simplifying, the reasoning is that we are good at intelligent tasks because we are also emotionally involved with them, so real AI will be achievable only if some 'emotional intelligence' can be developed. I hope you see this as a *modus tollens* (if *p* then *q*, but not *q*, therefore not *p*), but even if it is not, the premise that intelligence requires emotion seems to be in need of some justification. Vague evolutionary references and the usual anti-Cartesianism that is de rigueur are messy and confusing. There are plenty of intelligent animals that flourish without any ostensible reliance on emotions or feelings of any kind. Crocodiles do not cry and ants do not get annoyed with cicadas. A hot computer is one with a broken cooling system. Hopefully our ICTs won't be too emotional when we finally stop pampering them, as we have been forced to do for decades. It is high time for ICTs to grow up and move out of our mental space. This is the advantage of third-order technologies. Being left alone is the next big wave of innovations.

Finally, ACs will act as 'memory stewards', creating and managing a repository of information about their owners. This is good news. For leaving behind a lasting trace has always been a popular strategy to withstand the oblivion inevitably following one's death. Nowadays, we can all be slightly less forgettable, insofar as we succeed in our mnemonic DIY. This trend will grow exponentially, once ACs become commodities. We saw in Chapter 1 that storage capacity is increasing at an astonishing pace and at decreasing prices. Globally, it does not keep up with our production of even more data, but, locally, storage is certainly no longer a problem when it comes to the recording of a whole life by an AC. One day, Alice may receive her AC as a newborn, and keep it, upgrading it, repairing it, perhaps replacing it with new models, all her life, with her entire life archived in its files. Then, it will not be long before some smart application—based on a lifetime's recording of someone's voice, visual and auditory experiences, expressed opinions and tastes, linguistic habits, millions of digital documents, and so forth—will be able to *simulate* that person, to the point where one may interact with Alice's AC even after Alice's death, without noticing, or even deliberately disregarding, any significant difference. Early in 2011, some funeral homes in the US began attaching small Quick Response codes to gravestones, offering visitors the possibility to access information about the dead, such as online memorials, obituaries, or an interactive life story. The sky seems to be the right limit for such a business. A personalized AC could make one 'e-mortal'. After all, an advanced, customized ELIZA[11]—the famous program that used natural language processing to interact with users' responses on the basis of scripts—could already fool many people online (people have tried to date ELIZA-like agents online for years). Our new memory stewards will exacerbate old problems and pose new and difficult ones. What to erase, rather than what to record (as is already the case with one's emails), the safety and editing of what is recorded, the availability, accessibility, and transmission of the information recorded, its longevity, future consumption and 'replaying', the management of ACs that have outlived their human partners, the redressing of the fine balance between the art of forgetting and the process of forgiving (consider post-dictatorial, post-apartheid, or post-civil-war cultures), delicate issues in informational privacy, and the impact that all this will have on the construction of personal and social identities, and on the narratives that make up people's own past and roots: these are only some of the issues that will require careful handling, not only technologically, but also educationally and philosophically.

Artificial companions and many other ICTs' realizations of AI systems will increasingly interact with inforgs like us especially on the Web. So what the Web will look like in the near future is a question that has kept pundits and techno-fans on their toes for some time. The recent reshaping of the industry, with social media coming to maturity, has only increased the need for some clarity. In recent years, two distinct answers have gradually emerged from the rather vociferous and noisy market of ideas: one, unmistakeably Tim Berners-Lee's, advocates the Semantic Web; the other, easily recognizable as Tim O'Reilly's, supports the so-called Web 2.0. It makes a significant difference where semantic engines like us will be interacting with syntactic engines like our ICTs and Artificial Companions, so which Tim is right? The answer requires a new section.

The Semantic Web and its syntactic engines

Tim Berners-Lee introduced the idea of a Semantic Web in the nineties. Two decades or so later, it has become hard to disentangle a simple and clear definition of the Semantic Web, also known as Web 3.0, from the barrage of unrealistic and inflated hype or just unreliable and shameless advertisements. Let me provide a longish excerpt from one of the most quoted texts on the topic.[12] It is useful in order to illustrate some of the inflation in the idea of a Semantic Web.

> Most of the Web's content today is designed for humans to read, not for computer programs to manipulate *meaningfully*. Computers can adeptly parse Web pages for layout and routine processing—here a header, there a link to another page—but in general, computers have no reliable way to process the *semantics*.
>
> The Semantic Web will bring structure to the *meaningful content* of Web pages, creating an environment where software agents roaming from page to page can readily carry out sophisticated tasks for users. [...] it will 'know' all this without needing artificial intelligence on the scale of 2001's Hal or Star Wars's C-3PO.

The Semantic Web is not a separate Web but an extension of the current one, in which *information is given well-defined meaning*, better enabling computers and people to work in cooperation. [...] machines become much better able to process and '*understand*' the data that they merely display at present.

The Semantic Web will enable machines to *comprehend semantic documents* and data, not human speech and writings.

It all makes for fast-paced and exciting reading, full of promises. It is (or rather was, for a decade) representative of the literature on the Semantic Web. And yet, it is far from the more cautious and austere perspective endorsed by the World Wide Web Consortium (W3C), which describes the Semantic Web as (emphasis added)

A common framework that allows *data* to be shared and reused across application, enterprise, and community boundaries. [...] It is based on the Resource Description Framework (RDF).[13]

So who is right? And why the notable discrepancy?

Supporters of the Semantic Web may be understandably over-enthusiastic about its actual deliverability. It would be marvellous if we could build it. But unfortunately, a truly Semantic Web is an AI-complete problem[14] for which there is no foreseeable, technological solution. It is as nice and as unrealistic as *Star Wars*' C-3PO. Whereas a technically feasible, allegedly 'semantic' Web is unexciting, because it must necessarily fail to deliver what it promises, namely a Web in which computers *understand and interpret the meaning and significance* of the data they are processing. The truth is that a technically accurate description of a realistically feasible Semantic Web bears little resemblance to what one finds advertised. Let me quote the W3C once more (emphasis added):

The Semantic Web is a *web of data*. [...] The Semantic Web is about two things. It is about *common formats for integration and combination of data* drawn from diverse sources, where the original Web mainly concentrated on the interchange of documents. It is also about *language for recording how the data relates to real world objects*. That allows a person, or a machine, to start off in one database, and then move through an unending set of databases which are connected not by wires but by being about the same thing.

As the reader can see, it is *data* (not *semantic information,* which requires some *understanding*) and *syntax* (not *meaning,* which requires some *intelligence*) all the way through. We should really be speaking of the *Machine-readable Web* or indeed of the *Web of Data* as the W3C does. Such a *MetaSyntactic Web* works, and works increasingly well for circumscribed, standardized, and formulaic contexts, e.g., a catalogue of movies for online customers. This is really what the W3C is focusing on. Unexciting and, in its true colours, simply unsellable, which is a pity, because the MetaSyntactic Web is a genuinely useful development. Let us now look at Web 2.0.

Web 2.0 and its semantic engines

Providing a watertight definition of what qualifies as Web 2.0 might be an impossible rather than just a tricky task. But the fact that Web 2.0 refers to a loose gathering of a wide variety of family-resembling technologies, services, and products, is not a justification for a frustrating lack of clarity. A foggy environment is not a good reason for an out-of-focus picture of it. True, attempts to sharpen what we mean by Web 2.0 applications abound, but none of them has acquired the status of even a de facto standard. To be fair, Tim O'Reilly sought to be precise. In 2005, in a famous post entitled 'Web 2.0: Compact Definition?'[15] he wrote:

> Web 2.0 is the network as platform, spanning all connected devices; Web 2.0 applications are those that make the most of the intrinsic advantages of that platform: delivering software as a continually-updated service that gets better the more people use it, consuming and remixing data from multiple sources, including individual users, while providing their own data and services in a form that allows remixing by others, creating network effects through an 'architecture of participation,' and going beyond the page metaphor of Web 1.0 to deliver rich user experiences.

So the Semantic Web is really the *participatory Web,* which today includes 'classics' such as YouTube, eBay, Facebook, and so forth. Just check the top twenty-five websites in Alexa, the web service that provides information about websites.

So what is the difference between Web 2.0 and the Semantic Web? A good way to answer this question is by trying to understand the success of Web 2.0 applications in the last decade or so.

Web 2.0 works for the following reasons. Metadata are still data, even if about data, i.e. they are identifiable differences that only afford and constrain (but are still devoid of) semantic interpretation. They should not be confused with semantic information (which requires meaning), let alone knowledge (which requires truth and at least some form of explanation and understanding). However, our ICTs— including any smart artificial agent that can be actually built on the basis of our best understanding of current computer science—are *syntactic engines*, which cannot process meaning. So, the Semantic Web is largely mere hype: it is really based on *data* description languages; no semantic information is involved. On the contrary, humans are the only semantic engines available, the ghosts in the machines, as acknowledged by Twitter engineers. So Web 2.0 is the Web created by semantic engines for semantic engines, by relying on the contribution of legions of users. As an illustration, consider folksonomies.

A folksonomy (from folk and taxonomy) is the aggregated result of the social practice of producing information about other information (e.g., a photograph) through collaborative classification, known as social tagging (e.g., the photograph receives the tags 'New York', 'Winter', 'Statue of Liberty'). It works bottom-up, since it is left to the single individual user or producer of the tagged target to choose what to classify, how to classify it, and what appropriate keywords to use in the classification. Folksonomies have become popular since 2004, as an efficient way to personalize information and facilitate its fruition through information management tools. If you visit Flickr and search for 'New York', 'Winter', 'Statue of Liberty', you can retrieve the corresponding pictures of the Statute of Liberty in New York taken during the winter. Simple? Yes, but, it is trivial to object that folksonomies might be egregiously ambiguous. If you look further down in the list of photographs, you will find the picture of a person dressed like the Statue of Liberty, in New York, in the winter, and even the picture of a boat, called 'The Statue of Liberty'. The computer does not differentiate among these. It retrieves all the photographs that have been tagged 'New York', 'Winter', and 'Statue of Liberty'. Yet, this is not a problem for semantic engines like us, capable of fast disambiguation processes. We can often provide more tags as input (the photograph of the boat already had 69 tags at the time of writing), and spot the difference in the output anyway (it is difficult to confuse the statue with the person or the boat).

It turns out that Web 2.0 is an achievable and increasingly implemented reality, represented no longer by the creation of another, external space, like Web 1.0, but by an ecosystem friendly to, and inhabited by, humans as inforgs. Web 2.0 is a part of the infosphere where *memory as registration and timeless preservation* (the Platonic view) is replaced by *memory as accumulation and refinement*, and hence *search* replaces *recollection*. It is an environment characterized by its *time-friendliness*: time adds value and Web 2.0 applications and contents get better by use, that is, they improve with age, not least because the number of people involved is constantly increasing. This, in turn, is a function of a critical mass of inforgs who produce and consume semantic information. For example: with Wikipedia entries, the longer they are online and the more used the better, also because a whole new generation of an increasing number of participants escalates the peer-review effect. The objection that wikipedians should not be so dismissive of the *Britannica* or any other published source of information is correct since, after all, old, copyright-free entries from the *Britannica* are included in Wikipedia and this further supports its time-friendliness, since Wikipedia does get better precisely because it can easily cannibalize any other copyright-free resources available. Furthermore, the editorial structure of Wikipedia is far more complex, articulated, and 'hierarchical' than people normally seem to acknowledge. Self-generated contents are really the result of hard-driven and highly controlled processes. That anyone can contribute does not mean that anyone may. But this too, is time-friendly, since it relies on volunteers and their willingness to collaborate within an organization.

All this also helps to explain why it is preferable to see Web 2.0 today as part of cloud computing. This is another metaphor (and buzzword) for the Internet, also rather fuzzy and vague. However, as in the case of Web 2.0, cloud computing does capture a real new paradigm, when it is used to refer to an upgraded transformation of computing resources into utilities. Software tools, memory space, computational power, and other services or ICT-capabilities are all provided as Internet-based services (in the 'cloud') in a way that is entirely infrastructure-transparent and seamless to the user. It is the ultimate challenge to the spatial localization and hence fragmentation of information processes. Web 2.0 is time-friendly, cloud computing is space-friendly: it does not matter where you are but only what computational resources you need.

Web 1.0 and the Semantic Web are, on the contrary, time-unfriendly and fail to rely on the large number of small contributions that can be offered by millions of inforgs. For example, the longer a printed entry from the *Britannica* has been available the less useful it is likely to get, becoming utterly outdated in the long run. The same applies to old-fashioned websites working as hubs. So, a simple test to know whether something belongs to Web 2.0 is to ask: does it improve with time, usage, and hence number of people connected? We saw that services that pass the test include Flickr, YouTube, and Wikipedia.

Webs and the infosphere

The full Semantic Web is, I would contend, a well-defined mistake, whereas Web 2.0 is an ill-defined success. They are both interesting instances of the larger phenomena of *construction and defragmentation of the infosphere*. Web 2.0/the Participatory Web erases barriers between production and consumption of information (less friction) in one or more phases of the information life cycle (from occurrence through processing and management to usage, see Figure 4), or between pro-ducers and consumers of information. Web 3.0/the Semantic Web, understood, as it should be, as the MetaSyntactic Web, erases barriers between databases. We might then label Web 4.0 the Bridging Web, which erases the digital divide between who is and who is not a citizen of the information society (effective availability and accessibility), and historical vs. hyperhistorical societies. Interestingly, this is happening more in terms of smartphones and other hand-held devices—in Africa, China, and India—than in terms of a commodification of personal computers. According to a report by the International Tele-communication Union (ITU),[16] in 2013 there were 6.8 billion mobile subscriptions for 7.1 billion people, with the number of subscriptions surpassing that of people by early 2014. By Web 5.0, one may then refer to cloud computing and its ability to erase physical barriers and the global vs. local distinction. Finally, Web 6.0 would be the Web Onlife, which is erasing the threshold between the online and the offline worlds. These various Webs are developing in parallel and hence are only partially chronological in their order of appearance. Their numbering implies no hierarchical ordering; it is just a matter of convenient labelling. They should be seen more as converging forces pushing the evolution of the Web in the direction of a better info-sphere. Microsoft's 'input one' strategy, pursuing the development of a single device (such as the Xbox) that might represent the heart of our living rooms and our onlife experiences for all sorts of ICT applications, may be better understood in the light of such a unified infosphere.

The previous interpretation of the future of the Web—as develop-ing along the line of a progressive defragmentation of the space of information—outlines a broad scenario, according to which humans as social inforgs and semantic engines will inhabit an infosphere increasingly boundless, seamless, synchronized (time), delocalized (space), and correlated (interactions), to remind you about some of the qualifications we encountered in the previous chapters. It is an environment based on the gradual accrual and transmission of semantics through time by generations of inforgs, a collaborative effort to save and improve meaning for future refinement and reuse. This 'green policy' is the last point on which I would like to comment.

The reader may recall the disturbing scenes in *The Matrix* when we are finally shown batteries of humans farmed as mere biological sources of energy. It is a compelling story, but also an idiotic waste of resources. What makes humans special is not their bodies, which are not much better, and possibly worse, than the bodies many animals have, but the coalition of capacities which one may call intelligence or the mind. We could have tails, horns, wings, or plumes, be oviparous or live under the sea: the best use that one could still make of humanity as only a means and never as an end in itself, to mis-paraphrase Kant, would still be in terms of inforgs, organisms that are semantically omnivorous, capable of semantic processing and intelligent interactions. We generate and use meaning a bit like the larvae of the mulberry silkworm produce and use silk. It is an extra-ordinary feature, which so far appears unique in the universe, assuming that there are no other forms of advanced semantic intelligence like ours on other planets. It is also a feature that we have exploited only partially in the past. Civilizations, cultures, science, social trad-itions, languages, narratives, arts, music, poetry, philosophy... in short all the vast semantic input and output of billions of inforgs has been slowly layered for millennia like a thin stratum of humus on the hard bed of history. Too often it has been washed away by natural and man-made disasters, or made sterile by its inaccessibility or unavail-ability. Without it, human life is the life of a brute, of a mindless body. Yet the presence, preservation, accumulation, curation, expansion, and best use of semantics has been limited, when compared to what humanity has been able to achieve in the area of management of material and energy resources and the shaping of the physical envir-onment. The information revolution that we are experiencing today is partly understandable in terms of redressing such a lack of balance. ICTs have reached a stage when they might guarantee the stable presence, the steady accumulation and growth, and the increasing usability of our semantic humus. The good news is that building the infosphere as a friendly environment for future generations is becoming easier. The bad news is that, for the foreseeable future, the responsibility for such a gigantic task will remain totally human.

Conclusion

Light AI, smart agents, artificial companions, Semantic Web, or Web 2.0 applications are part of what I have described as a fourth revolution in the long process of reassessing humanity's fundamental nature and role in the universe. The deepest philosophical issue brought about by ICTs concerns not so much how they extend or empower us, or what they enable us to do, but more profoundly how they lead us to reinterpret who we are and how we should interact with each other. When artificial agents, including artificial companions and software-based smart systems, become commodities as ordinary as cars, we shall accept this new conceptual revolution with much less reluctance. It is humbling, but also exciting. For in view of this important evolution in our self-understanding, and given the sort of ICT-mediated interactions that humans will increasingly enjoy with other agents, whether natural or synthetic, we have the unique opportunity of developing a new ecological approach to the whole of reality. As I shall argue in Chapter 10, how we build, shape, and regulate ecologically our new infosphere and ourselves is the crucial challenge brought about by ICTs and the fourth revolution.

Recall Beatrice's question at the beginning of *Much Ado About Nothing*: 'Who is his companion now?' She would not have understood 'an artificial agent' as an answer to her question. I suspect future generations will find it unproblematic. It is going to be our task to ensure that the transition from her question to their answer will be as acceptable as possible. Such a task is both ethical and political and, as you may expect by now, this is the topic of Chapter 8.

8

POLITICS

The Rise of the Multi-Agent Systems

Political apoptosis

We saw in Chapter 1 that history has lasted 6,000 years, since it began with the invention of writing in the 4th millennium BC. During this relatively short time, ICTs provided the *recording* and *transmitting* infrastructure that made the escalation of other technologies possible. This gradually increased our dependence on more and more layers of technologies. ICTs entered into a more mature phase in the few centuries between Gutenberg and Turing. Today, their autonomous *processing* capacities have ushered in a new, hyperhistorical age. Information societies depend on first-, second-, and third-order ICTs for societal welfare, personal well-being, technological innovation, scientific discoveries, and economic growth. Some data may help to bring home the point more clearly.

In 2011, the total world wealth[1] was calculated to be $231 trillion, up from $195 trillion in 2010.[2] Since we are almost 7 billion, that was about $33,000 per person, or $51,000 per adult, as the report indicates. The figures give a clear sense of the level of inequality. In the same year, we spent $498 billion on advertisements.[3] Perhaps for the first time, we also spent more on ways to entertain ourselves than on ways to kill each other. The military expenditure in 2010 was $1.74 trillion,[4] and that on entertainment and media was expected to be around

$2 trillion, with digital entertainment and media share growing to 33.9 per cent of all spending by 2015, from 26 per cent in 2011.[5] Meanwhile, we spent $6.5 trillion (this is based on 2010 data) on fighting health problems and premature death, much more than the military and the entertainment and media budgets put together. All these trillions were closely linked and often overlapped with the budget for ICTs, on which we spent $3 trillion in 2010.[6] We can no longer unplug our world from ICTs without turning it off.

If the analysis offered in the previous chapters is even approximately correct, humanity's emergence from its historical age represents one of the most significant steps it has ever taken. It certainly opens up a vast horizon of opportunities as well as challenges and difficulties, all essentially driven by the recording, transmitting, and processing powers of ICTs. From synthetic biochemistry to neuroscience, from the Internet of things to unmanned planetary explorations, from green technologies to new medical treatments, from social media to digital games, from agricultural to financial applications, from economic developments to the energy industry, our activities of discovery, invention, design, control, education, work, socialization, entertainment, care, security, business, and so forth would be not only unfeasible but unthinkable in a purely mechanical, historical context. They have all become hyperhistorical in nature.

Hyperhistory, and the evolution of the infosphere in which we live, are quickly detaching future generations from ours. Of course, this is not to say that there is no continuity, both backwards and forwards. *Backwards*, because it is often the case that the deeper a transformation is, the longer and more widely rooted its causes may be. It is only because many different forces have been building the pressure for a long time that radical changes may happen all of a sudden, perhaps unexpectedly. It is not the last snowflake that breaks the branch of the tree. In our case, it is certainly history that begets hyperhistory. There is no ASCII (American Standard Code for Information Interchange) without the alphabet. *Forwards*, because we should expect historical societies to survive for a long time in the future, not unlike the prehistoric Amazonian tribes mentioned in Chapter 1. Despite global-ization, human societies do not parade uniformly forward, in neat and synchronized steps.

We are witnessing a slow and gradual process of political *apoptosis*. Apoptosis, also known as programmed cell death, is a natural and normal form of self-destruction in which a programmed sequence of events leads to the self-elimination of cells. Apoptosis plays a crucial role in developing and maintaining the health of the body. One may see this as a natural process of renovation. Here, I am using the expression 'political apoptosis' in order to describe the gradual and natural process of renovation of sovereign states[7] as they develop into information societies (see Figure 21). Let me explain.

Simplifying and generalizing, a quick sketch of the last 400 years of political history in the Western world may look like this. The Peace of Westphalia (1648) meant the end of World War Zero, namely the Thirty Years War, the Eighty Years War, and a long period of other conflicts during which European powers, and the parts of the world they controlled, massacred each other for economic, political, and religious reasons. Christians brought hell to each other, with staggering violence and unspeakable horrors. The new system that emerged in those years, the so-called *Westphalian order*, saw the coming maturity of sovereign states and then national states as we still know them today: France, for example. Think of the time between the last chapter of *The Three Musketeers*—when D'Artagnan, Aramis, Porthos, and Athos take part in Cardinal Richelieu's siege of La Rochelle in 1628—and the first chapter of *Twenty Years Later*, when they come together again, under the regency of Queen Anne of Austria (1601–66) and the rule of Cardinal Mazarin (1602–61).

The state did not become a monolithic, single-minded, well-coord-inated entity. It was not the sort of beast that Hobbes described in his *Leviathan*, nor the sort of robot that a later, mechanical age would incline us to imagine. But it did rise to the role of the binding power, the system able to keep together and influence all the different agents comprising it, and coordinate their behaviours, as long as they fell within the scope of its geographical borders. These acquired the metaphorical status of a state's skin. States became the independent agents that played the institutional role in a system of international relations. And the principles of sovereignty (each state has the fundamental right of political self-determination), legal equality (all states are equal), and non-intervention (no state should interfere with the internal affairs of another state) became the foundations of such a system of international relations.

Citizenship had been discussed in terms of biology (your parents, your gender, your age . . .) since the early city states of ancient Greece. It became more flexible (types of citizenship) when it was conceptualized in terms of legal status as well. This was the case under the Roman Empire, when acquiring citizenship—a meaningless idea in purely biological contexts—meant becoming a holder of rights. With the modern state, geography started playing an equally important role, mixing citizenship with language, nationality, ethnicity, and locality. In this sense, the history of the passport is enlightening. As a means to prove the holder's identity, it is acknowledged to be an invention of King Henry V of England (1386–1422), a long time before the Westphalian order took place. However, it was the Westphalian order that transformed the passport into a document that entitles the holder not to travel (because a visa may also be required, for example) or be protected abroad, but to return to the country that issued the passport. The passport became like an elastic band that ties the holder to a geographical point, no matter how far in space and prolonged in time the journey in other lands has been. Such a document became increasingly useful the better that geographical point was defined. Travelling was still quite passport-free in Europe until the First World War. Only then did security pressure and techno-bureaucratic means catch up with the need to disentangle and manage all those elastic bands travelling around by means of a new network, the railway.

Back to the Westphalian order. Now that the physical and legal spaces overlap, they can both be governed by sovereign powers, which exercise control, impose laws, and ensure their respect by means of physical force within the state's borders. Geographical mapping is not just a matter of travelling and doing business, but also an inward-looking question of controlling one's own territory, and an outward-looking question of positioning oneself on the globe. The taxman and the general look at those geographical lines with eyes very different from those of today's users of Expedia. For sovereign states act as agents that can, for example, raise taxes within their borders and contract debts as legal entities (hence our current terminology in terms of 'sovereign bonds', for example, which are bonds issued by national governments in foreign currencies), and of course dispute borders, often violently. Part of the political struggle becomes not just a silent tension between different components of the state as a multi-agent system, say the clergy vs. the aristocracy, but an explicitly codified balance between the different agents constituting it. In particular, Montesquieu (1689–1755) suggested the classic division of the state's political powers that we take for granted today: a legislature, an executive, and a judiciary. The state as a multi-agent system organizes itself as a network of these three 'small worlds', among which only some specific channels of information are allowed. Today, we may call that arrangement Westphalian 2.0.

With the Westphalian order, modern history becomes the age of the state. The state arises as *the* information agent, which legislates on, and at least tries to control, the technological means involved in the information life-cycle, including education, census,[8] taxes, police records, written laws, press, and intelligence. Already most of the adventures in which D'Artagnan is involved are caused by some secret communication.

As the information agent, the state fosters the development of ICTs as a means to exercise and maintain legal force, political power, and social control, especially at times of international conflicts, frequent unrests, and fragile peace. For example, in 1790–5, during the French Revolution, the French government needed a system of speedy communication to receive intelligence and transmit orders in time to counterbalance the hostile manoeuvres of the allied forces that surrounded France: Britain, the Netherlands, Prussia, Austria, and Spain. To satisfy such need, Claude Chappe (1763–1805) invented the first system of telegraphy (he actually coined the word 'telegraph'). It consisted of mechanical semaphores that could transmit messages across the country in a matter of hours. It became so strategic that when Napolen begun preparations to resume war in Italy in 1805, he ordered a new extension from Lyons to Milan. At its peak, the Chappe telegraph was a network of 534 stations, covering more than 5,000 km (3,106 miles). The reader may remember its crucial appearance in Alexandre Dumas's *The Count of Monte Cristo* (1844) where the Count bribes an operator to send a false message to manipulate the financial market to his own advantage. In fictional as in real life, whoever controls information controls the issuing events.

Through the centuries, the state moves from being conceived as the ultimate guarantor and defender of a laissez-faire society to a Bismarckian welfare system, which takes full care of its citizens. In both cases, the state remains the primary collector, producer, and controller of information. However, by fostering the development of ICTs, the state ends by undermining its own future as the only, or even the main, information agent. This is the political apoptosis I mentioned earlier. For in the long run, ICTs contribute to transforming the state in an information society, which makes possible other, sometimes even more powerful, information agents, which may determine political decisions and events. And so ICTs help shift the balance against centralized government, in favour of distributed governance and international, global coordination.

The two world wars are also clashes of sovereign states resisting mutual coordination and inclusion as part of larger multi-agent systems. The Bretton Woods conference may be interpreted as the event that seals the beginning of the political apoptosis of the state. The gathering in 1944 of 730 delegates from all 44 Allied nations at the Mount Washington Hotel in Bretton Woods, New Hampshire, United States, regulated the international monetary and financial order after the conclusion of Second World War. It saw the birth of the

International Bank for Reconstruction and Development (this, together with the International Development Association, is now known as the World Bank), of the General Agreement on Tariffs and Trade (GATT, replaced by the World Trade Organization in 1995), and the International Monetary Fund. In short, Bretton Woods brought about a variety of multi-agent systems as supranational or intergovernmental forces involved with the world's political, social, and economic problems. These and similar agents became increasingly powerful and influential, as the emergence of the Washington Consensus clearly indicated.

John Williamson[9] coined the expression 'Washington Consensus' in 1989. He used it in order to refer to a set of ten specific policy recommendations, which, he argued, constituted a standard strategy adopted and promoted by institutions based in Washington, DC— such as the US Treasury Department, the International Monetary Fund, and the World Bank—when dealing with countries coping with economic crises. The policies concerned macroeconomic stabilization, economic opening with respect to both trade and investment, and the expansion of market forces within the domestic economy. In the past quarter of a century, the topic has been the subject of intense and lively debate, in terms of correct description and acceptable prescription. Like the theory of a Westphalian doctrine I outlined earlier, the theory of a Washington Consensus is not devoid of problems. Does the Washington Consensus capture a real historical phenomenon? Does the Washington Consensus ever achieve its goals? Is it to be reinterpreted, despite Williamson's quite clear definition, as the imposition of neo-liberal policies by Washington-based international financial institutions on troubled countries? These are important questions, but the real point of interest here is not the interpretative, economic, or normative evaluation of the Washington Consensus. Rather, it is the fact that the very idea, even if it remains only an influential idea, captures a significant aspect of our hyperhistorical, post-Westphalian time. The Washington Consensus is a coherent development of Bretton Woods. Both highlight the fact that, after the Second World War, organizations and institutions (not only those in Washington DC) that are not states but rather non-governmental multi-agent systems, are openly acknowledged to act as major influential forces on the political and economic scene internationally, dealing with global problems through global policies. The very fact that the Washington Consensus has been accused (no matter whether correctly or not) of disregarding local specificities and global differences reinforces the point that a variety of powerful multi-agent systems are now the new sources of policies in the globalized information societies. As a final reminder, let me mention a rather controversial report, entitled *Top 200: The Rise of Corporate Global Power*. It offered some years ago an analysis of corporate agents.[10] Perhaps the most criticized part was a comparison between countries' yearly GDP and companies' yearly sales (revenues or turnover). Despite this potential shortcoming, it still makes for interesting reading. According to the report:

> of the 100 largest economies in the world, 51 are [as of 2000] corporations; only 49 are [as of 2000] countries.

The criticism remains, but the percentage has probably moved in favour of the number of companies, and what represents a unifying unit of comparison is that both GDP and revenues buy you clout. When multi-agent systems of such dimensions take decisions, their effects are deep and global.

Today, we know that global problems—from the environment to the financial crisis, from social justice to intolerant religious fundamentalisms, from peace to health conditions—cannot rely on sovereign states as the only source of a solution because they involve and require global agents. However, there is much uncertainty about the design of the new multi-agent systems that may shape humanity's future. Hyperhistorical societies are post-Westphalian, because of the emergence of the sovereign state as the modern political-information agent. They are post-Bretton Woods, because of the emergence of non-state multi-agent systems as hyperhistorical players in the global

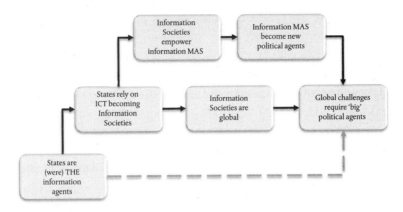

Fig. 21. The Emergence of Political Multi-Agent Systems (MAS).

economy and politics. This helps explain why one of the main challenges faced by hyperhistorical societies is how to design the right sort of multi-agent systems. These systems should take full advantage of the sociopolitical progress made in modern history, while dealing successfully with the new global problems, which undermine the legacy of that very progress, in hyperhistory.

A new informational order?

The shift from a historical, Westphalian order to a post-Bretton Woods, hyperhistorical predicament in search of a new equilibrium may be explained by many factors. Four are worth highlighting in the context of this book.

First, *power*. We saw that ICTs 'democratize' data and the processing/controlling power over them, in the sense that now both tend to reside and multiply in a multitude of repositories and sources. Thus, ICTs can create, enable, and empower a potentially boundless number of non-state agents, from the single individual to associations and groups, from macro-agents, like multinationals, to international, intergovernmental as well as non-governmental, organizations and supranational institutions. The state is no longer the only, and sometimes not even the main, agent in the political arena that can exercise informational power over other informational agents, in particular over human individuals and groups. The European Commission, for example, recognized the importance of such new agents in the *Cotonou Agreement* between the European Union (EU) and the African, Caribbean, and Pacific (ACP) countries, by acknowledging the important role exercised by a wide range of non-governmental development actors, and formally recognizing their participation in ACP–EU development cooperation. According to article 6 of the Cotonou Agreement, such non-state actors comprise

> the private sector; economic and social partners, including trade union organisations; civil society in all its forms, according to national characteristics.

The 'democratization' brought about by ICTs is generating a new tension between power and force, where power is informational, and exercised through the elaboration and dissemination of norms, whereas force is physical, and exercised when power fails to orient the behaviour of the relevant agents and norms need to be enforced. Note that the more physical goods and even money become information-dependent, the more the informational power exercised by multi-agent systems acquires a significant financial aspect.

Second, *geography*. ICTs de-territorialize human experience. They have made regional borders porous or, in some cases, entirely irrelevant. They have also created, and are exponentially expanding, regions of the infosphere where an increasing number of agents, not necessarily only human, operate and spend more and more time: the onlife experience. Such regions are intrinsically stateless. This is generating a new tension between geopolitics, which is global and non-territorial, and the state, which still defines its identity and political legitimacy in terms of a sovereign territorial unit, as a country.

Third, *organization*. ICTs fluidify the topology of politics. They do not merely enable but actually promote, through management and empowerment, the agile, temporary, and timely aggregation, disaggregation, and reaggregation of distributed groups 'on demand', around shared interests, across old, rigid boundaries, represented by social classes, political parties, ethnicity, language barriers, physical barriers, and so forth. This is generating new tensions between the state, still understood as a major organizational institution, yet no longer rigid but increasingly morphing into a flexible multi-agent system itself, and a variety of equally powerful, indeed sometimes even more powerful and politically influential (with respect to the old sovereign state), non-state organizations, the other multi-agent systems on the block. Terrorism, for example, is no longer just a problem concerning internal affairs—consider forms of terrorism in the Basque Country, Germany, Italy, or Northern Ireland—but also an international confrontation with a distributed multi-agent system such as al-Qaeda.

Finally, *democracy*. Changes in power, geography, and organization reshape the debate on democracy, the oldest and safest form of power crowdsourcing. We used to think that, ideally, democracy should be a direct and constant involvement of all citizens in the running of their society and its business, their *res publica*. Direct democracy, if feasible, was about how the state could reorganize itself internally, by designing rules and managing the means to promote forms of negotiation, in which citizens could propose and vote on policy initiatives directly and almost in real time. We thought of forms of direct democracy as complementary options for forms of representative democracy. It was going to be a world of 'politics always-on'. The reality is that direct democracy has turned into a mass-media-led democracy, in the ICT sense of new social media. In such digital democracies, distributed groups, temporary and timely aggregated around shared interests, have multiplied and become sources of influence external to the state. Citizens vote for their representatives but can constantly influence them via opinion polls almost in real time. Consensus-building has become a constant concern based on synchronic information.

Because of the factors just analysed—power, geography, organization, and democracy—the unique position of the historical state as *the* information agent is being undermined from below and overridden from above. Other multi-agent systems have the data, the power, and sometimes even the force—as in the different cases of the UN, of groups' cyber threats, or of terrorist attacks—the space, and the organizational flexibility to erode the modern state's political clout. They can appropriate some of its authority and, in the long run, make it redundant in contexts where it was once the only or the predominant informational agent. The Greek economic crisis, which began in late 2009, offers a good example. The Greek government and the Greek state had to interact 'above' with the EU, the European Central Bank, the International Monetary Fund, the rating agencies, and so forth. They had to interact 'below' with the Greek mass media and the people in Syntagma Square, the financial markets and international investors, German public opinion, and so forth. Because the state is less central than in the nineteenth century, countries such as Belgium and Italy may work fine even during long periods without governments or when governed by dysfunctional ones, on 'automatic pilot'.

A much more networked idea of political interactions makes possible a degree of tolerance towards, and indeed feasibility of, localisms and separatisms, as well as movements and parties favouring autonomy or independence that would have been unacceptable by modern nation states, which tended to encourage aggregating forms of nationalism but not regionalism. From Padania (Italy) to Catalonia (Spain), from Scotland (Great Britain) to Bavaria (Germany), one is reminded that, in almost any European country, hyperhistorical trends may resemble pre-Westphalian equilibria among a myriad of regions. The long 'list of active separatist movements in Europe' in Wikipedia is both informative and eye-opening. Unsurprisingly, the Assembly of European Regions (originally founded as the Council of the Regions of Europe in 1985), which brings together over 250 regions from 35 countries along with 16 interregional organizations, has long been a supporter of subsidiarity, the decentralizing principle according to which political matters ought to be dealt with by the smallest, lowest, or least centralized authority that could address them effectively.

Of course, the historical state is not giving up its role without a fight. In many contexts, it is trying to reclaim its primacy as the information super-agent governing the political life of the society that it organizes.

In some cases, the attempt is blatant. In the UK, the Labour Government introduced the first Identity Cards Bill in November 2004. After several intermediary stages, the Identity Cards Act was finally repealed by the Identity Documents Act 2010, on 21 January 2011. The failed plan to introduce compulsory ID in the UK should be read from a modern perspective of preserving a Westphalian order.

In many other cases, it is 'historical resistance' by stealth, as when an information society is largely run by the state. In this case, the state maintains its role of major informational agent no longer just legally, on the basis of its power over legislation and its implementation, but also economically, on the basis of its power over the majority of information-based jobs. The intrusive presence of so-called State Capitalism with its State-Owned Enterprises all over the world, from Brazil, to France, to China, is a symptom of hyperhistorical anachronism.

Similar forms of resistance seem only able to delay the inevitable rise of political multi-agent systems. Unfortunately, they may involve not only costs, but also huge risks, both locally and globally. Recall that the two world wars may be seen as the end of the Westphalian system. Paradoxically, while humanity is moving into a hyperhistorical age, the world is witnessing the rise of China, currently a most 'historical' state, and the decline of the US, a state that more than any other superpower in the past already had a hyperhistorical and multi-agent vocation in its federal organization. We might be moving from a Washington Consensus to a *Beijing Consensus* described by Williamson as consisting of incremental reform, innovation and experimentation, export-led growth, state capitalism, and authoritarianism.[11] This is risky, because the anachronistic historicism of some of China's policies and humanity's growing hyperhistoricism are heading towards a confrontation. It may not be a conflict, but hyperhistory is a force whose time has come, and while it seems likely that it will be the Chinese state that will emerge deeply transformed, one can only hope that the inevitable friction will be as painless and peaceful as possible. The financial and social crises that the most advanced information societies are undergoing may actually be the painful but still peaceful price we need to pay to adapt to a future post-Westphalian system.

The previous conclusion holds true for the historical state in general. In the future, political multi-agent systems will acquire increasing prominence, with the problem that the visibility and transparency of such acquisition of power may be rather unclear. It is already difficult to monitor and understand politics when states are the main players. It becomes even harder when the agents in question have fuzzier features, more opaque behaviours, and are much less easily identifiable, let alone accountable. At the same time, it is to be hoped that the state itself will progressively abandon its resistance to hyperhistorical changes and evolve even more into a multi-agent system. Good examples are provided by devolution, the transfer of a state's sovereign rights to supranational European institutions, or the growing trend in making central banks, like the Bank of England or the European Central Bank, independent, public organizations.

The time has come to consider the nature of a political multi-agent system more closely and some of the questions that its emergence is already posing.

The political multi-agent system

A political multi-agent system is a single agent, constituted by other systems, which is

teleological: the multi-agent system has a purpose, or goal, which it pursues through its actions;

interactive: the multi-agent system and its environment can act upon each other;

autonomous: the multi-agent system can change its configurations without direct response to interaction, by performing internal transformations to change its states—this imbues the multi-agent system with some degree of complexity and independence from its environment—and finally

adaptable: the multi-agent system's interactions can change the rules by which the multi-agent system itself changes its states. Adaptability ensures that the multi-agent system learns its own mode of operation in a way that depends critically on its experience.

The political multi-agent system becomes *intelligent* (in the AI sense discussed in Chapter 6) when it implements the previous features efficiently and effectively, minimizing resources, wastefulness, and errors, while maximizing the returns of its actions.

The emergence of intelligent, political multi-agent systems poses many serious questions. Some of them are worth reviewing here, even if only quickly: identity, cohesion, consent, social vs. political space, legitimacy, and transparency.

Identity. Throughout modernity, the state has dealt with the problem of establishing and maintaining its own identity by working on the equation between state and nation. This has often been achieved through the legal means of citizenship and the narrative rhetoric of space (the mother/fatherland) and time (story in the sense of traditions, recurrent celebrations of past nation-building events, etc.). Consider, for example, the invention of mandatory military service during the French Revolution, its increasing popularity in modern history, but then the decreasing number of sovereign states that still impose it nowadays (your author belongs to the last generation that had to serve in the Italian army for twelve months). Conscription transformed waging war from an eminently economic problem—Florentine bankers financed the English kings during the Hundred Years War (1337–1453), for example—into also a legal problem: the right of the state to send its citizens to die on its behalf. It thus made human life the *penultimate* value, available for the ultimate sacrifice, in the name of patriotism: 'for King and Country'. It is a sign of modern anachronism that, in moments of crisis, sovereign states still give in to the temptation of fuelling nationalism about meaningless *geographical* spots, often some small islands unworthy of any human loss, including the Falkland Islands (UK) or Islas Malvinas (Argentina), the Senkaku (Japan) or Diaoyu (China) islands, and the Liancourt Rocks, also known as Dokdo (South Korea) or Takeshima (Japan).

Cohesion. The equation between state and nation, achieved through citizenship and land/story, had the further advantage of providing an answer to a second problem, that of cohesion. For the equation answered not only the question of who or what the state is, but also the question of who or what belongs to the state and hence may be subject to its norms, policies, and actions. New political multi-agent systems cannot rely on the same solution. Indeed, they face the further problem of having to deal with the decoupling of their political identity and cohesion. The political identity of a multi-agent system may be strong and yet unrelated to its temporary and rather loose cohesion, as is the case with the Tea Party movement in the US. Both identity and cohesion of a political multi-agent system may be rather weak, as in the international Occupy movement. Or one may recognize a strong cohesion and yet an unclear or weak political identity, as with the population of tweeting individuals and their role during the Arab Spring. Both identity and cohesion of a political multi-agent system are established and maintained through information sharing. The land is virtualized into the region of the infosphere in which the multi-agent system operates. So memory (retrievable recordings) and coherence (reliable updates) of the information flow enable a political multi-agent system to claim some identity and some cohesion, and therefore offer a sense of belonging. But it is, above all, the fact that the boundaries between the online and offline are disappearing, the appearance of the onlife experience, and hence the fact that the virtual infosphere can affect politically the physical space, that reinforces the sense of the political multi-agent system as a real agent. If *Anonymous* had only a virtual existence, its identity and cohesion would be much less strong. Deeds provide a vital counterpart to the virtual information flow to guarantee cohesion. Interactions become more fundamental than things, in a way that is coherent with what we have seen in Chapters 2 (interactability as a criterion of existence) and 3 (informational identity). With wordplay, we might say that *ings* (as in interact-*ing*, process-*ing*, network-*ing*, do-*ing*, be-*ing*, etc.) replace *things*.

Consent. The breaking-up of the equation 'political multi-agent system = sovereign state, citizenship, land, story, nation' and the decoupling of identity and cohesion in a political multi-agent system have a significant consequence. The age-old theoretical problem of how consent to be governed by a political authority arises is being turned on its head. In the historical framework of social contract theory, the presumed default position is that of a legal opt-out. There is some kind (to be specified) of original consent, allegedly given (for a variety of reasons) by any individual subject to the political state, to be governed by the latter and its laws. The problem is to understand how such consent is given and what happens when an agent, especially a citizen, opts out of it (the outlaw). In the hyperhistorical framework, the expected default position is that of a social opt-in, which is exercised whenever the agent subjects itself to the political multi-agent system conditionally, for a specific purpose. Simplifying, we are moving from being part of the political consensus to taking part in it, and such part-taking is increasingly 'just in time', 'on demand', 'goal-oriented', and anything but stable, permanent, or long-term. If doing politics looks increasingly like doing business this is because, in both cases, the interlocutor, the citizen-customer, needs to be convinced to behave in a preferred way every time anew. Loyal membership is not the default position, and needs to be built and renewed around political and commercial products alike. Gathering consent around specific political issues becomes a continuous process of (re-)engagement. It is not a matter of limited attention span. The generic complaint that 'new generations' cannot pay sustained attention to political problems any more is ill-founded. They are, after all, the generations that binge-watch TV. It is a matter of motivating interest again and again, without running into an inflation of information (one more crisis, one more emergency, one more revolution, one more...) and political fatigue (how many times do we need to intervene urgently?). Therefore, the problem is to understand what may motivate repeatedly or indeed force agents (again, not just individual human beings, but all kinds of agents) to give such consent and become engaged, and what happens when such agents, unengaged by default (note, not disengaged, for disengagement presupposes a previous state of engagement), prefer to stay away from the activities of the political multi-agent system, inhabiting a social sphere of civil but apolitical identity.

Failing to grasp the previous transformation from historical opt-out to hyperhistorical opt-in means being less likely to understand the apparent inconsistency between the disenchantment of individuals with politics and the popularity of global movements, international mobilizations, activism, voluntarism, and other social forces with huge political implications.[12] What is moribund is not politics *tout court*, but historical politics, that based on parties, classes, fixed social roles, political manifestos and programmes, and the sovereign state, which sought political legitimacy only once and spent it until revoked. The inching towards the so-called centre by parties in liberal democracies around the world, as well as the 'get out the vote' strategies (the expression is used to describe the mobilization of *voters as supporters* to ensure that those who can, do vote), are evidence that engagement needs to be constantly renewed and expanded in order to win an election. Party (as well as union) membership is a modern feature that is likely to become increasingly less common.

Social vs. Political Space. In prehistory, the social and the political spaces overlap because, in a stateless society, there is no real difference between social and political relations and hence interactions. In history, the state tends to maintain such coextensiveness by occupying, as an informational multi-agent system, the social space politically, thus establishing the primacy of the political over the social. This trend, if unchecked and unbalanced, risks leading to totalitarianisms (consider for example the Italy of Mussolini), or at least broken democracies (consider next the Italy of Berlusconi). We have seen earlier that such coextensiveness and its control may be based on normative or economic strategies, through the exercise of power, force, and rule-making. In hyperhistory, the social space is the original, default space from which agents may move to (consent to) join the political space. It is not accidental that concepts such as *civil society*,[13] *public sphere*,[14] and *community* become increasingly important the more we move into a hyperhistorical context. The problem is to understand and design such social space where agents of various kinds are supposed to be interacting and which give rise to the political multi-agent system.

Each agent within the social space has some degrees of freedom. By this I do not mean liberty, autonomy, or self-determination, but rather, in the robotic, more humble sense, some capacities or abilities, supported by the relevant resources, to engage in specific actions for a specific purpose. To use an elementary example, a coffee machine has only one degree of freedom: it can make coffee, once the right ingredients and energy are supplied. The sum of an agent's degrees of freedom is its 'agency'. When the agent is alone, there is of course only agency, but no social, let alone political, space. Imagine Robinson Crusoe on his 'Island of Despair'. However, as soon as there is another agent (Friday on the 'Island of Despair'), or indeed a group of agents (the native cannibals, the shipwrecked Spaniards, the English mutineers), agency acquires the further value of social interaction. Practices and then rules for coordination and constraint of the agents' degrees of freedom become essential, initially for the well-being of the agents constituting the multi-agent system, and then for the well-being of the multi-agent system itself. Note the shift in the level of analysis: once the social space arises, we begin to consider the group as a group— e.g., as a family, or a community, or as a society—and the actions of the individual agents constituting it become elements that lead to the newly established degrees of freedom, or agency, of the multi-agent system. The previous simple example may still help. Consider now a coffee machine and a timer: separately, they are two agents with different agency, but if they are properly joined and coordinated into a multi-agent system, then the issuing agent has the new agency to make coffee at a set time. It is now the multi-agent system that has a more complex capacity, and that may or may not work properly.

A social space is the totality of degrees of freedom of the inhabiting agents one wishes to take into consideration. In history, such consideration—which is really just another level of analysis—was largely determined physically and geographically, in terms of presence in a territory, and hence by a variety of forms of neighbourhood. In the previous example, all the agents interacting with Robinson Crusoe are taken into consideration because of their relations (interactive presence in terms of their degrees of freedom) to the same 'Island of Despair'. We saw that ICTs have changed all this. In hyperhistory, where to draw the line to include, or indeed exclude, the relevant agents whose degrees of freedom constitute the social space has become increasingly a matter of at least implicit choice, when not of explicit decision. The result is that the phenomenon of distributed morality, encompassing that of distributed responsibility, is becoming more and more common. In either case, history or hyperhistory, what counts as a social space may be a political move. Globalization is a de-territorialization in this political sense.

Turning now to the political space in which the new multi-agent systems operate, it would be a mistake to consider it a separate space, over and above the social one. Both the social and the political space are determined by the same totality of the agents' degrees of freedom. The political space emerges when the complexity of the social space requires the prevention or resolution of potential *divergences* and coordination or collaboration about potential *convergences*. Both are crucial. And in each case information is required, in terms of representation and deliberation about a complex multitude of degrees of freedom.

Legitimacy. It is when the agents in the social space agree to agree on how to deal with their divergences (conflicts) and convergences that the social space acquires the political dimension to which we are so used. Yet two potential mistakes await us here.

The first, call it Hobbesian, is to consider politics merely as the prevention of war by other means, to invert the famous phrase by

Clausewitz (1780–1831), according to whom 'war is the continuation of politics by other means'. This is an unsatisfactory view of politics, because even a complex society of angels would still require rules in order to further its harmony. Convergences too need politics. Politics is not just about conflicts due to the agents' exercises of their degrees of freedom when pursuing their goals. It is also, or at least it should be, above all, the furthering of coordination and collaboration of degrees of freedom by means other than coercion and violence.

The second potential mistake, which may be called Rousseauian, is to misunderstand the political space as just that part of the social space organized by law. In this case, the mistake is subtler. We usually associate the political space with the rules or laws that regulate it but the latter are not constitutive, by themselves, of the political space. Compare two cases in which rules determine a game. In chess, the rules do not merely constrain the game; they are the game because they do not supervene on a previous activity. Rather, they are the *necessary and sufficient conditions* that determine all—and the only— moves that can be legally made. In football, however, the rules are supervening *constraints* because the agents enjoy a previous and basic degree of freedom, consisting in their capacity to kick a ball with the foot in order to score a goal, which the rules are supposed to regulate. Whereas it is physically possible, but makes no sense, to place two pawns on the same square of a chessboard, nothing impeded Maradona from scoring an infamous goal by using his hand in the Argentina vs. England football match (1986 FIFA World Cup), and that to be allowed by a referee who did not see the infringement. Now, the political space is not simply *constituted* by the laws that regulate it, as in the chess example. But it is not just the result of the *constraining* of the social space by means of laws either, as in the football example. The political space is that area of the social space *configured* by the agreement to agree on resolution of divergences and coordination of convergences. The analogy here is the formatting of a hard disk. This leads to a further consideration, concerning the transparent multi-agent system, especially when, in this transition time, the multi-agent system in question is still the state.

Transparency. There are two senses in which the multi-agent system can be transparent. They mean quite different things, and so they can be confusing. Unsurprisingly, both come from ICTs and computer science, one more case in which the information revolution is changing our conceptual framework.

On the one hand, the multi-agent system (think of the sovereign state, and also of corporate agents, multinationals, or supranational institutions, etc.) can be transparent in the sense that it moves from being a black box to being a white box. Other agents (citizens, when the multi-agent system is the state) not only can see inputs and outputs—for example, levels of tax revenue and public expenditure—they can also monitor how (in our running example, the state as) a multi-agent system works internally. This is not a novelty at all. It was a principle already popularized in the 19th century. However, it has become a renewed feature of contemporary politics due to the possibilities opened up by ICTs. This kind of transparency is also known as *Open Government.*

On the other hand, and this is the more innovative sense that I wish to stress here, the multi-agent system can be transparent in the sense of being 'invisible'. This is the sense in which a technology (especially an interface) is transparent: not because it is not there, but because it delivers its services so efficiently, effectively, and reliably that its presence is imperceptible. When something works at its best, behind the scenes as it were, to make sure that we can operate as smoothly as possible, then we have a transparent system. When the multi-agent system in question is the state, this second sense of transparency should not be seen as a surreptitious way of introducing, with a different terminology, the concept of 'small state' or 'small governance'. On the contrary, in this second sense, the multi-agent system (the state) is as transparent and as vital as the oxygen that we breathe. It strives to be the ideal butler. There is no standard terminology for this kind of transparent multi-agent system that becomes perceivable only when it is absent. Perhaps one may speak of *Gentle Government*.

It seems that multi-agent systems can increasingly support the right sort of ethical infrastructure (more on this later) the more transparently, that is, openly and gently, they play the negotiating game through which they take care of the *res publica*. When this negotiating game fails, the possible outcome is an increasingly violent conflict among the parties involved. It is a tragic possibility that ICTs have seriously reshaped.

All this is not to say that *opacity* does not have its virtues. Care should be exercised, lest the sociopolitical discourse is reduced to the nuances of higher quantity, quality, intelligibility, and usability of information and ICTs. The more the better is not the only, nor always the best, rule of thumb. For the withdrawal of information can often make a positive and significant difference. We already encountered Montesquieu's division of the state's political powers. Each of them may be informationally opaque in the right way to the other two. For one may need to lack (or intentionally preclude oneself from accessing) some information in order to achieve desirable goals, such as protecting anonymity, enhancing fair treatment, or implementing unbiased evaluation. Famously, Rawls's 'veil of ignorance' exploits precisely this aspect of information, in order to develop an impartial approach to justice.[15] Being informed is not always a blessing and might even be dangerous or wrong, distracting or crippling. The point about the value of transparency is that its opposite, informational opacity, cannot be assumed to be a good property of a political system unless it is adopted explicitly and consciously, by showing that it is a feature not a mere bug.

Infraethics

Part of the ethical efforts engendered by the fourth revolution concerns the design of environments that can facilitate ethical choices, actions, or process. This is not the same as *ethics by design*. It is rather *pro-ethical design*, as I hope will become clearer in what follows. Both are liberal, but *ethics by design* may be mildly paternalistic, insofar as it privileges the facilitation of the *right* kind of choices, actions, process, or interactions on behalf of the agents involved. Whereas *pro-ethical design* does not have to be paternalistic, insofar as it privileges the facilitation of *reflection* by the agents involved on their choices, actions, or process. For example, strategies based on *ethics by design* may let you opt out of the *default* preference according to which, by obtaining a driving licence, you are also willing to be an organ donor. Strategies based on *pro-ethical design* may not allow you to obtain a driving licence unless you have indicated whether you wish to be an organ donor: the unbiased choice is still all yours. In this section, I shall call environments that can facilitate ethical choices, actions, or process, the ethical infrastructure, or *infraethics*. The problem is how to design the right sort of infraethics. Clearly, in different cases, the design of a liberal infraethics may be more or less paternalistic. My argument is that it should be as little paternalistic as the circumstances permit, although no less.

It is a sign of the times that, when politicians speak of infrastructure nowadays, they often have in mind ICTs. They are not wrong. From business fortunes to conflicts, what makes contemporary societies work depends increasingly on bits rather than atoms. We already saw all this. What is less obvious, and intellectually more interesting, is that ICTs seem to have unveiled a new sort of ethical equation.

Consider the unprecedented emphasis that ICTs have placed on crucial phenomena such as trust, privacy, transparency, freedom of expression, openness, intellectual property rights, loyalty, respect, reliability, reputation, rule of law, and so forth. These are probably better understood in terms of an infrastructure that is there to facilitate or hinder (reflection upon) the im/moral behaviour of the agents involved. Thus, by placing our informational interactions at the centre of our lives, ICTs seem to have uncovered something that, of course, has always been there, but less visibly so: the fact that the moral behaviour of a society of agents is also a matter of 'ethical infrastructure' or simply infraethics. An important aspect of our moral lives has escaped much of our attention. Many concepts and related phenomena have been mistakenly treated as if they were only ethical, when in fact they are probably mostly infraethical. To use a term from the philosophy of technology, such concepts and the corresponding phenomena have a dual-use nature: they can be morally good, but also morally evil (more on this presently). The new equation indicates that, in the same way that, in an economically mature society, business and administration systems increasingly require infrastructures (transport, communication, services etc.) to prosper, so too, in an informationally mature society, multi-agent systems' moral interactions increasingly require an infraethics to flourish.

The idea of an infraethics is simple, but can be misleading. The previous equation helps to clarify it. When economists and political scientists speak of a 'failed state', they may refer to the failure of a *state-as-a-structure* to fulfil its basic roles, such as exercising control over its borders, collecting taxes, enforcing laws, administering justice, providing schooling, and so forth. In other words, the state fails to provide *public goods*, such as defence and police, and *merit goods*, such as health care. Or (too often an inclusive and intertwined or) they may refer to the collapse of a *state-as-an-infrastructure* or environment, which makes possible and fosters the right sort of social interactions. This means that they may be referring to the collapse of a substratum of default expectations about economic, political, and social conditions, such as the rule of law, respect for civil rights, a sense of political community, civilized dialogue among differently minded people, ways to reach peaceful resolutions of ethnic, religious, or cultural tensions, and so forth. All these expectations, attitudes, practices—in short such an implicit 'sociopolitical infrastructure', which one may take for granted—provide a vital ingredient for the success of any complex society. They play a crucial role in human interactions, comparable to the one that we are now accustomed to attributing to physical infrastructures in economics.

Infraethics should not be understood in terms of Marxist theory, as if it were a mere update of the old 'base and superstructure' idea. The elements in question are entirely different: we are dealing with moral actions and not-yet-moral facilitators of such moral actions. Nor should it be understood in terms of a kind of second-order normative discourse on ethics. It is the not-yet-ethical framework of implicit expectations, attitudes, and practices that *can* facilitate and promote moral decisions and actions. At the same time, it would also be wrong to think that an infraethics is morally neutral. Rather, it has a dual-use nature, as I anticipated earlier: it can both facilitate and hinder morally good as well as evil actions, and do this in different degrees. At its best, it is the grease that lubricates the moral mechanism. This is more likely to happen whenever having a 'dual-use' nature does not mean that each use is equally likely, that is, that the infraethics in question is still not neutral, nor merely positive, but does have a bias to deliver more good than evil. If this is confusing, think of the dual-use nature not in terms of a state of equilibrium, like an ideal coin that can deliver both heads and tails, but in terms of a co-presence of two alternative outcomes, one of which is more likely than the other, as a biased coin more likely to turn heads than tails. When an infraethics has a 'biased dual-use' nature, it is easy to mistake the infraethical for the ethical, since whatever helps goodness to flourish or evil to take root partakes of their nature.

Any successful complex society, be it the City of Man or the City of God, relies on an implicit infraethics. This is dangerous, because the increasing importance of an infraethics may lead to the following risk: that the legitimization of the ethical discourse is based on the 'value' of the infraethics that is supposed to support it. *Supporting* is mistaken for *grounding*, and may even aspire to the role of *legitimizing*, leading to what the French philosopher Jean-François Lyotard (1924–98) criticized as mere 'performativity' of the system, independently of the actual values cherished and pursued. As an example, think of a bureaucratic context in which some procedure, supposed to deliver some morally good behaviour, through time becomes a value in itself, and ends giving ethical value to the behaviour that it was supposed to support. Infraethics is the vital syntax of a society, but it is not its semantics, to reuse a distinction we encountered when discussing artificial intelligence. It is about the structural form, not the meaningful contents.

We saw earlier that even a society in which the entire population consisted of angels, that is, perfectly moral agents, still needs norms for collaboration and coordination. Theoretically, a society may exist in which the entire population consisted of Nazi fanatics who could rely on high levels of trust, respect, reliability, loyalty, privacy, transparency, and even freedom of expression, openness, and fair competition. Clearly, what we want is not just the successful mechanism provided by the right infraethics, but also the coherent combination between it and morally good values, such as civil and political rights. This is why a balance between security and privacy, for example, is so difficult to achieve, unless we clarify first whether we are dealing with a tension within ethics (security and privacy as moral rights), within infraethics (both are understood as not-yet-ethical facilitators), or between infraethics (security) and ethics (privacy), as I suspect. To rely on another analogy: the best pipes (infraethics) may improve the flow but do not improve the quality of the water (ethics); and water of the highest quality is wasted if the pipes are rusty or leaky. So creating the right sort of infraethics and maintaining it is one of the crucial challenges of our time, because an infraethics is not morally good in itself, but it is what is most likely to yield moral goodness if properly designed and combined with the right moral values. The right sort of infraethics should be there to support the right sort of values. It is certainly a constitutive part of the problem concerning the design of the right multi-agent systems.

The more complex a society becomes, the more important and hence salient the role of a well-designed infraethics is, and yet this is exactly what we seem to be missing. Consider the recent Anti-Counterfeiting Trade Agreement (ACTA), a multinational treaty concerning the international standards for intellectual property rights.[16] By focusing on the enforcement of intellectual property rights (IPR), supporters of ACTA completely failed to perceive that it would have undermined the very infraethics that they hoped to foster, namely one promoting some of the best and most successful aspects of our information society. It would have promoted the structural inhibition of some of the most important individuals' positive liberties and their ability to participate in the information society, thus fulfilling their own potential as informational organisms. For lack of a better word, ACTA would have promoted a form of *informism*, comparable to other forms of inhibition of social agency such as classism, racism, and sexism. Sometimes a protection of liberalism may be inadvertently illiberal. If we want to do better, we need to grasp that issues such as IPR are part of the new infraethics for the information society, that their protection needs to find its carefully balanced place within a complex legal and ethical infrastructure that is already in place and constantly evolving, and that such a system must be put at the service of the right values and moral behaviours. This means finding a compromise, at the level of a liberal infraethics, between those who see new legislation (such as ACTA) as a simple fulfilment of existing ethical and legal obligations (in this case from trade agreements), and those who see it as a fundamental erosion of existing ethical and legal civil liberties.

In hyperhistorical societies, any regulation affecting how people deal with information is now bound to influence the whole infosphere and onlife habitat within which they live. So enforcing rights such as IPR becomes an environmental problem. This does not mean that any legislation is necessarily negative. The lesson here is one about complexity: since rights such as IPR are part of our infraethics and affect our whole environment understood as the infosphere, the intended and unintended consequences of their enforcement are widespread, interrelated, and far-reaching. These consequences need to be carefully considered, because mistakes will generate huge problems that will have cascading costs for future generations, both ethically and economically. The best way to deal with 'known unknowns' and unintended consequences is to be careful, stay alert, monitor the development of the actions undertaken, and be ready to revise one's decision and strategy quickly, as soon as the wrong sort of effects start appearing. *Festina lente*, 'more haste, less speed', as the classic adage suggests. There is no perfect legislation but only legislation that can be perfected more or less easily. Good agreements about how to shape our infraethics should include clauses about their timely updating.

Finally, it is a mistake to think that we are like outsiders ruling over an environment different from the one we inhabit. Legal documents (such as ACTA) emerge from within the infosphere that they affect. We are building, restoring, and refurbishing the house from inside. Recall that we are repairing the raft while navigating on it, to use the metaphor introduced in the Preface. Precisely because the whole problem of respect, infringement, and enforcement of rights such as IPR is an infraethical and environmental problem for advanced information societies, the best thing we can do, in order to devise the right solution, is to apply to the process itself the very infraethical framework and ethical values that we would like to see promoted by it. This means that the infosphere should regulate itself from within, not from an impossible without.

Hyperhistorical conflicts and cyberwar

The story goes that when the Roman horsemen first saw Pyrrhus' 20 war elephants, at the Battle of Heraclea (280 BC), they were so terrorized by these strange creatures, which they had never seen before, that they galloped away, and the Roman legions lost the battle. Today, the new elephants are digital. The phenomenon might have just begun to emerge in the public debate but, in hyperhistorical societies, ICTs are increasingly shaping armed conflicts.

Disputes become armed conflicts when politics fails. In hyperhistory, such armed conflicts have acquired a new informational nature. Cyberwar or information warfare is the continuation, and sometimes the replacement, of conflict by digital means, to rely once more on

Clausewitz's famous interpretation of war we encountered earlier. Four main changes are notable.

First, in terms of conventional military operations, ICTs have progressively revolutionized communications, making possible complex new modes of field operations. We saw this was already the case with the Chappe telegraph.

Second, ICTs have also made possible the swift analysis of vast amounts of data, enabling the military, intelligence, and law enforcement communities to take action in ever more timely and targeted ways. ICTs and big data are also weapons.

Third, and even more significantly, battles are nowadays fought by highly mobile forces, armed with real-time ICT devices, satellites, battlefield sensors, and so forth, as well as thousands of robots of all kinds.

And, finally, the growing dependence of societies and their militaries on advanced ICTs has led to strategic cyber-attacks, designed to cause costly and crippling disruption. Armies of human soldiers may no longer be needed. This creates a stark contrast with suicide terrorism. On the one hand, human life can regain its ultimate value because the state no longer needs to trump it in favour of patriotism. Contrary to what we saw earlier, drones do not die 'for King and Country'. Cyberwar is a hyperhistorical phenomenon. On the other hand, terrorists dehumanize individuals as mere delivery mechanisms. Suicide terrorism is a historical phenomenon, in which the technology in-between is the human body and a person becomes a 'living tool', using Aristotle's definition of a slave encountered in Chapter 2.

The old economic problem—how to finance war and its expensive high-tech—is now joined by a new legal problem: how to reconcile a hyperhistorical kind of warfare with historical phenomena, such as the infringement of national sovereignty and respect for geographical borders. Furthermore, cyber attacks can be undertaken by nations or networks, or even by small groups or individuals. ICTs have made asymmetric conflicts easier, and shifted the battleground more than an inch into the infosphere.

The scale of such transformations is staggering. For example, in 2003, at the beginning of the war in Iraq, US forces had no robotic systems on the ground. However, by 2004, they had already deployed 150 robots, in 2005 the number was 2,400, and by the end of 2008 about 12,000 robots of nearly two dozen varieties were operating on the ground.[17]

In 2010, Neelie Kroes, vice-president of the European Commission, commenting on Cyber Europe 2010, the first pan-European cyber-attack simulation, said:

> This exercise to test Europe's preparedness against cyber threats is an important first step towards working together to combat potential online threats to essential infrastructure and ensuring citizens and businesses feel safe and secure online.[18]

As you can see, the perspective could not be more hyperhistorical.

ICT-mediated modes of conflict pose a variety of ethical problems, for war-fighting militaries in the field, for intelligence-gathering services, for policymakers, and for ethicists. They may be summarized as the three Rs: risks, rights, and responsibilities.

Risks. Cyberwar and information-based conflicts may increase risks, making 'soft' conflicts more likely and hence potentially increasing the number of casualties. Between 2004 and 2012, drones operated by the US Central Intelligence Agency (CIA) killed more than 2,400 people in Pakistan, including 479 civilians, with 3 strikes in 2005 escalating to 76 strikes in 2011.[19] A troubling perspective is that ICTs might make unconventional conflicts more acceptable ethically, by stressing the less deadly outcome of military operations in cyberspace. However, this might be utterly illusory. Messing with ICT infrastructures of hospitals and airports may easily cause the loss of human lives, even if in a less obvious way than bombs do. Despite this, the mistaken impression remains that we might be allegedly moving towards a more precise, surgical, bloodless way of handling violently our political disagreements.

Rights. Cyberwar tends to erase the threshold between reality and simulation, between life and play, and between conventional conflicts, insurgencies, or terrorist actions. This threatens to increase the potential tensions between fundamental rights: informational threats require higher levels of control, which may generate conflicts between individuals' rights (e.g. privacy) and community's rights (e.g. safety and security). A state's duty to protect its citizens may come to clash with its duty to prevent harm to its citizens, via an extended system of surveillance, which may easily end up infringing on citizens' privacy.

Responsibilities. Cyberwar makes it more difficult to identify responsibilities that are reshaped and distributed. Because causal links are much less easily identifiable, it becomes much more difficult to establish who, or what, is accountable and responsible when software/ robotic weapons and hybrid, man-machine systems are involved.

New risks, rights, and responsibilities: in short, cyberwar is a new phenomenon, which has caught us by surprise. With hindsight, we should have known better, for at least three reasons.

Take the nature of our society first. When it was modern and industrial, conflicts had mechanized, second-order features. Engines, from battleships to tanks to aeroplanes, were weapons, and the coherent outcome was the emphasis on energy, petrol first and then nuclear power. There was an eerie analogy between assembly lines and warfare trenches, between working force and fighting force. Conventional warfare was kinetic warfare. We just did not know it, because the non-kinetic kind was not yet available. The Cold War and the emergence of asymmetric conflicts were part of a post-industrial transformation. Today, in a culture in which we have seen that the word 'engine' is more likely to be preceded by the verb 'search' than by the noun 'petrol', hyperhistorical societies are as likely to fight with digits as they are with bullets, with computers as well as guns, not least because digital systems tend to be in charge of analogue weapons. I am not referring to the use of intelligence, espionage, or cryptography, but to cyber attacks or the extensive use of drones and other military robots in Iraq and Afghanistan. It is old news. On 27 April 2007, about

1 million computers worldwide were used for DDOS (distributed denial of service) attacks on Estonian government and corporate websites. A DDOS attack is a systematic attempt to make computer resources unavailable, at least temporarily, by forcing vital sites or services to reset or consume their resources, or by disrupting their communications so that they can no longer function properly. Russia was blamed but denied any involvement. In June 2010, Stuxnet, a sophisticated computer malware, sabotaged c.1,000 Siemens centrifuges used in the Iranian nuclear power plant of Bushehr. That time, the US and Israel denied any involvement. At the time of writing, there is an ongoing attack on US ICT infrastructure. This time it is China that denies any involvement. Then there are robotic weapons, which may be seen as the final stage in the industrialization of warfare, or, more interestingly, as the first step in the development of information conflicts, in which command and control as well as action and reaction become tele-concepts. Third-order technological conflicts in which humans are no longer in the loop have moved out of science fiction and into military scenarios. From software agents in cyberspace to robots in physical environments, we should not be too optimistic about the non-violent nature of cyberwar. The more we rely on ICTs, the more we envelop the world, the more cyber attacks will become lethal. Soon, crippling an enemy's communication and information infrastructure will be like zapping its pacemaker rather than hacking its mobile.

Second, consider the nature of our environment. We have been talking about the Internet and cyberspace for decades. We could have easily imagined that this would become the new frontier for human conflicts. Technologies have continuously expanded. We have been fighting each other on land, at sea, in the air, and in space for as long, and as soon, as technologies made it possible. Predictably, the infosphere was never going to be an exception. Information is the fifth element,[20] and the military now speaks of cyberwarfare as 'the fifth domain of warfare'. The impression is that, in the future, such a fifth domain will end up dominating the others. The following two examples may help. On 13 May 1999, arguably the first combat between an aircraft and an unmanned drone took place when an Iraqi MiG-25 shot down a US Air Force unmanned MQ-1 Predator drone. More than 360 drones have been built since 1995, for more than $2.38 billion. Second, since 2006, Samsung (the maker of the smart refrigerator we met in Chapter 2) has also been producing the SGR-A1. It is a robot with a low-light camera and pattern recognition software to distinguish humans from animals or other objects. It patrols South Korea's border with North Korea and, if necessary, it can autonomously fire its built-in machine gun. It is increasingly hard to draw a clear distinction between cyberwarfare and conventional, kinetic warfare when some tele-warfare is in question.

Finally, think of the origin of cybernetics, the computer, the Internet, the Global Positioning System (GPS), and unmanned drones and vehicles. They all developed initially as part of wider military efforts. The history of computing is deeply rooted in the Second World War and Turing's work at Bletchley Park. Cybernetics, the ancestor of contemporary robotics, began to develop as an engineering field in connection with applications for the automatic control of gun mounts and radar antenna, also during the Second World War. We know that the Internet was the outcome of the arms race and of nuclear proliferation, but we were distracted by the development of the Web and its scientific origins, and forgot about the Defense Advanced Research Projects Agency (DARPA). The now ubiquitous GPS, which provides the satellite-based information for navigation systems, was created and developed by the US Department of Defense, one more case of the political importance of geography. It became freely available for civilian use only in 1983, after a Boeing 747 of Korean Airlines, with 269 people on board, was shot down because it had strayed into the USSR's prohibited airspace. Finally, the development of drones, mainly but not only by the US military, as well as autonomous vehicles (DARPA again) and other robots, owes much to the conflicts in Iraq and Afghanistan and the fight against terrorism. In short, much of the history of digital ICTs spookily corresponds to the history of conflicts and the financial efforts behind them: Second World War,

Cold War, First and Second Iraq War, War in Afghanistan, and various 'wars' on terrorist organizations around the world. Hyperhistory has merely caught up with us.

The previous outline should help one understand why cyberwar, or more generally information warfare, is causing radical transformations in our ways of thinking about military, political, and ethical issues. The concepts of state, war, and the distinction between civil society and military organizations are being affected. Are we going to see a new arms race, given the high rate at which cyber weapons 'decay'? After all, you can use a piece of malware only once, for a patch will then become available, and often only within, and against, a specific technology that will soon be out of date. If cyber disarmament is ever going to be an option, how do you decommission cyber weapons? Digital systems can be hacked: will the Pony Express make a patriotic comeback in the near future as the last line of defence against an enemy that could tamper with anything digital and online? Some questions make one smile, but others are increasingly problematic. Let me highlight two sets of them that should be of more general interest.

The body of knowledge and discussion behind Just War Theory is detailed and extensive.[21] It is the result of centuries of refinements since Roman times. The methodological question we face today is whether information warfare is merely one more area of application, or whether it represents a disruptive novelty as well, which will require new developments of the theory itself. For example, within the *jus ad bellum*, which kind of authorities possess the legitimacy to wage cyberwar? And how should a cyber attack be considered in terms of last resort, especially when a cyber attack could, allegedly, prevent more violent outcomes? And within the *jus in bello*, what level of proportionality should be attributed to a cyber attack? How do you surrender to cyber enemies, especially when their identities are unknown on purpose? Or how will robots deal with non-combatants or treat prisoners? Is it possible or even desirable to develop inbuilt 'ethical algorithms' when engineering robotic weapons?

Equally developed, in this case since Greek times, is our understanding of military virtue ethics. How is the latter going to be applied to phenomena that are actually reshaping the conditions of possibility of virtue ethics itself? Bear in mind that any virtue ethics presupposes a philosophical anthropology, that is, a view of the human nature that may be Aristotelian, Buddhist, Christian, Confucian, Fascist, Nietzschean, Spartan, and so forth. And we saw in the previous chapters that information warfare is only part of the information revolution, which is also affecting our self-understanding as informational organisms. Take for example the classic virtue of courage: in what sense can someone be courageous when tele-manoeuvring a military robot? Indeed, will courage still rank so highly among the virtues when the capacity to evaluate and manage information and act upon it wisely and promptly will seem to be a much more important trait of a soldier's character?

Similar questions seem to invite new theorizing, rather than the mere application or adaptation of old ideas. ICTs have caused radical changes both in how societies may come into conflict and how they may manage it. At the same time, there is a policy and a conceptual deficit. For example, the US Department of Defense intends to replace a third of its armed vehicles and weaponry with robots by 2015, but it still lacks an ethical code for the deployment of these new, semi-autonomous weapons.[22] This is a global issue. The 2002 Prague Summit marked NATO's first attempt to address cyber-defence activities. Five years later, in 2007, there were already 42 countries working on military robotics, including Iran, China, Belarus, and Pakistan,[23] but not even a draft of an international agreement regarding their ethical deployment. There is a serious need for more descriptive and conceptual analyses of such a crucial area in applied ethics, and more assessment of the effectiveness of the initial measures that have been taken to deal with the increasing application of ICTs in armed conflicts. The issue could not be more pressing and there is a much felt and quickly escalating need to share information and coordinate ethical theorizing. The goals should be sharing information and views about the current state of the ethics of information warfare, developing a comprehensive framework for a clear interpretation of the new aspects of cyberwar, building a critical consensus about the ethical deployment of e-weapons, and laying down the foundation for an ethical approach to information warfare. We experimented with chemical weapons, especially during the First World War, and with biological weapons, in particular during the Sino-Japanese War of 1931–45. The horrific results led, in 1925, to the Geneva Protocol, prohibiting the use of chemical and biological weapons. In 1972, the Biological and Toxin Weapons Convention (BWC) banned the development, production, and storage of bioweapons. Since then, we have managed to restrain their use and, by and large, respect the BWC. Something similar happened with nuclear weapons. The hope is that information warfare and e-weapons will soon be equally regulated and constrained, without having to undergo any terrible and tragic lesson.

Let us return to the elephants. During the civil war, in the Battle of Thapsus (46 BC), Julius Caesar's fifth legion was armed with axes and was ordered to strike at the legs of the enemy's elephants. The legion withstood the charge, and the elephant became its symbol. Interestingly, nobody at the time could even imagine that there might be an ethical problem in treating animals so cruelly. We should think ahead, because history occasionally is a bit petulant and likes to repeat itself. At a time when there is an exponential growth in R & D concerning ICT-based weapons and strategies, we should collaborate on the identification, discussion, and resolution of the unprecedented ethical difficulties characterizing cyberwar. This is far from being premature. Perhaps, instead of updating our old ethical theories with more and more service packs, we might want to consider upgrading them by developing new ideas. Like the civilian uses of robots that we encountered in Chapter 6, information warfare calls for an information ethics. After all, iRobot produces both the Roomba 700 that vacuum-cleans your floor and the iRobot 710 Warrior that disposes of your enemies' explosives.

Conclusion

Six thousand years ago, humanity witnessed the invention of writing and the emergence of the conditions of possibility that lead to cities, kingdoms, empires, sovereign states, nations, and intergovernmental organizations. This is not accidental. Prehistoric societies are both ICT-less and stateless. The state is a typical historical phenomenon. It emerges when human groups stop living a hand-to-mouth existence in small communities and begin to live a mouth-to-hand one. Large communities become political societies, with division of labour and specialized roles, organized under some form of government, which manages resources through the control of ICTs, including that special kind of information called 'money'. From taxes to legislation, from the administration of justice to military force, from census to social infrastructure, the state was for a long time the ultimate information agent and so history, and especially modernity, is the age of the state.

Almost halfway between the beginning of history and now, Plato was still trying to make sense of both radical changes: the encoding of memories through written symbols and the symbiotic interactions between the individual and the *polis*-state. In fifty years, our grandchildren may look at us as the last of the historical, state-organized generations, not so differently from the way we look at the Amazonian tribes mentioned in Chapter 1, as the last of the prehistorical, stateless societies. It may take a long while before we come to understand in full such transformations. And this is a problem, because we do not have another six millennia in front of us. We are playing a technological gambit with ICTs, and we have only a short time to win the environmental game, for the future of our planet is at stake, as I shall argue in Chapter 9.

9

ENVIRONMENT

The Digital Gambit

The costs and risks of the anthropocene

The shift from history to hyperhistory, the construction and interpretation of the infosphere, living onlife, inscribing and enveloping the world: these are all immense changes that require unimaginable quantities of energy. Like a demiurge, a god that does not create but shapes the universe, humanity is modifying a whole planet to fit and satisfy its needs, wishes, and expectations. In Chapter 1, we saw that we have been doing this for millennia. Exactly how many is a matter of scientific dispute, but there is much agreement among geologists that the significant impact we have had on the Earth's ecosystems needs to be recognized by adopting a new, formal unit of geological epoch divisions, the 'anthropocene'.[1]

So far, the 'anthropocene' seems to have been a fairly successful story. However, this success has come at increasingly high environmental costs, some of which have recently become unsustainable. The development of the infosphere is now jeopardizing the well-being of the biosphere. This is a risk that is inevitable, but it should certainly be managed more safely, and it could be managed entirely successfully, as I shall argue presently.

Technologies lower constraints and expand opportunities. By doing so, they continuously redesign the feasibility space of the agents who

enjoy them, increasing their degrees of freedom. The more empowering or enabling technologies become, the more likely they are to
change the nature and scope of the risks that they may bring, both
in terms of undesirable outcomes (possible damages or losses) and in
terms of missed desirable outcomes (potential benefits and advantages, and what economists call opportunity costs). As a consequence,
technologies, by their nature, also tend to redesign the corresponding
space of risks in which agents operate and interact. As when making a
string of paper dolls, it seems that technologies cannot shape actual
constraints and opportunities without also shaping the corresponding
risks, both negative (missed desirable outcomes) and positive (potential undesirable outcomes). Therefore, a risk-free technology is an
oxymoron, as disasters and crises affecting the energy industry keep
painfully reminding us. Nonetheless, the intrinsically risky nature of
technologies should not be reason for despair. For technologies can
also reduce the space of risks and make it more manageable, and this is
ground for some cautious optimism. Let me explain.

Through time, the rather simple cycle of decreased constraints,
increased opportunities, and the corresponding new risks transforms
the set of risk-takers into a subset of the much larger set of risk-
runners or stakeholders. Given a specific risk, all risk-takers (those
who choose a risky option) are risk-runners (those affected by the risk
being taken), but there are many more risk-runners who are not also
the risk-takers. Already the identification of a chariot driver as the only
taker *and* runner of the same relevant risks seems implausible, to the
extent that, even in ancient Rome, there were laws regulating traffic
and driving behaviour. Such complete overlap between risk-taker and
risk-runner becomes inconceivable once we consider a taxi driver.
Now, in politically organized societies, the risk-runners seek to protect
themselves from the consequences of the actions of the risk-takers
through systems of regulations about standards, protocols, licences,
controls, deployment conditions, proper use, safety measures, insurances, and so forth. Once such regulations become formalized into
legislation, risk management can rely on legal systems and safety

technologies in order to establish constraints and provide opportunities in the development or use of a technology, while minimizing the risks involved. They both work in the same direction and both can be seen as part of the solution. Together, legal systems and safety technologies constitute what may be called *metatechnologies*. These are the kind of second- or third-order technologies that operate on (rather than with) and regulate other technologies.

The idea that some metatechnologies might be used to implement the safe, effective, and economical use of other technologies is not new. It was already theorized in the eighties, following the Three Mile Island accident, a partial nuclear meltdown that occurred in Pennsylvania, on 28 March 1979. Indeed, one could argue that the first governor, designed by James Watt (1736–1819) in 1788, was already a classic example of a second-order metatechnology used to regulate the amount of fuel used by an engine and hence maintain its speed as constant as possible. However, what I have in mind here is something slightly more inclusive. It is the view that a metatechnology should be understood as comprising not only the relevant technologies that regulate the appropriate technologies to which they apply, but also the rules, conventions, laws, and in general the sociopolitical conditions that regulate technological R & D and the following use or application of technologies. It is this broad concept of metatechnology that provides the aforementioned ground for some cautious optimism, in the following sense.

Consider potential negative risks first, the missed desirable outcomes of a technology. Much of the information economy has been made possible by ICTs also working as metatechnologies, enabling agents to identify benefits and exploit opportunities. Likewise, as a metatechnology, legislation can deal with negative risks by offering incentives to agents to become potential risk-takers. Germany is a good example. Because of its solar subsidy, Germany's solar energy market is by far the biggest in the world, with a total installed capacity in 2011 equivalent to 40 per cent of the world's total. Admittedly, known risks are a bit like pain: they might be unwelcome, but they often signal the presence of some significant trouble. So incentives, like painkillers, should be dispensed with care, as they might have serious counter-effects, in terms of hiding old problems, delaying their solutions, or causing new ones. In the case of solar panels, Germany amended its feed-in tariff law in 2012 in order to slow down the exponential growth in installations because of the financial costs and distorting effects on the energy market. Likewise, from an environmental perspective, it is important to recall that five of the top ten solar-panel makers in the world are from China, which dominated 40 per cent of the world's market in 2010. Unfortunately, Chinese industry has been repeatedly criticized for its poor record on working conditions, human rights, and environmental concerns. A similar analysis could be extended to many other cases, such as the sustainability of corn-based ethanol, coal-to-liquid synthetic fuels, or gas extracted through hydraulic fracturing (fracking). And yet, all this should not make us despair. If carefully handled, incentives may turn into wise investments, and build the essential bridge that is required if the energy industry is to transition from polluting to cleaner and renewable sources. That the path might be narrow does not mean that it is not worth pursuing. That it might be the only path merely reinforces the urgency of taking the right steps.

Consider next the potential positive risks, the undesirable outcomes of a technology. A metatechnological legislative approach is often at its best not when it provides affordances by offering incentives to counterbalance negative risks, but when it imposes constraints by enforcing disincentives to cope with positive risks, that is, when it focuses on the don'ts rather than the dos. In this case, the path is much broader and is represented by four main strategies: prevention, limitation, repair, and compensation.

Prevention. Once again, no metatechnological strategy is infallible. Prevention may be too radical when it imposes a complete ban on a particular technology. For example, in the 1970s, Italy, one of the earliest adopters of nuclear energy, was the third largest producer in the world, but a referendum in 1987, immediately after the Chernobyl disaster (1986), resulted in the phasing-out of all existing plants, causing an increasing dependence on energy imports and subsequent electricity prices much higher than the EU average. Unsurprisingly, the country was reconsidering the possibility of building nuclear plants, when the Fukushima Daiichi nuclear disaster in 2011 halted the political process (more on this presently). After all, Italy buys electricity from neighbouring France, which produces almost 80 per cent of its electric power through its nuclear network.

Limitation and *repair.* Relative prevention may be understood as measures that allow a technology to develop but seek to prevent, or at least limit, the realization of its risks, a bit like the ABS (anti-locking brakes). Prevention and, when it fails, limitation and repair of unwanted outcomes that have actually occurred, are metatechnological strategies that allow for degrees. This means that they may be more carefully tuned. The more flexible such strategies are, the more they tend to rely on the correct coordination between relevant legislation and safety technological solutions. However, both may still fail. For example, on 11 March 2011, following the Tōhoku earthquake and tsunami, the 16 nuclear plants in Japan affected by the earthquake were switched off within two minutes (including Fukushima), with cooling procedures initiating immediately and correctly. It was evidence of remarkable resilience that almost all nuclear plants could survive such a major natural disaster undamaged. Following current legislation, the Fukushima plant was protected by a sea wall designed to withstand a wave 5.7 m (19 ft) high, but the wave that struck it was about 14 m (46 ft) high and easily flooded the generator building. The ensuing problems and hazards were a consequence of a failure of the safety as much as of the legal metatechnological systems. In 2012, a Japanese parliamentary panel concluded that the crisis at the Fukushima nuclear plant was 'a profoundly man-made disaster'.[2] In all these cases, however, the crucial point is to realize that increasingly unwanted outcomes require ever more—not fewer—advanced, forward-looking, and sophisticated kinds of metatechnologies (legislation and safety technologies). This is where good design can make a significant difference, both by decreasing the chances that unwanted outcomes might occur, and by incorporating high degrees of resilience that can attenuate the effects of such unwanted outcomes when they do occur. The second worst thing, after a system's failure, is a system incapable of coping with its failure successfully.

Compensation. This fourth metatechnological strategy may also be unsuccessful, if badly designed. Compensations are not strategies to cope with the unwanted outcomes of a technology, in the same sense in which home insurance is not a way of coping with fire hazards. They should not be seen as deterrents either. If that is the goal, then legislation should probably ban the technology in question, or set up a fine and a points system in which the relevant international authority, for example the International Atomic Energy Agency, is empowered to issue fines and demerits to agents for the losses or damages caused by their technological mistakes. Compare this to some countries' legal systems whereby a driver's licensing authority issues demerits to drivers convicted for road traffic offences. Compensations are a means to manage the costs before (insurance premium) or after (repayment) a technology fails. They too may be less than effective, if not carefully calibrated. In 1990, after the Exxon Valdez spill in Alaska, the US Congress passed the Oil Pollution Act in order to make holders of leases or permits for offshore facilities liable for up to $75 million per spill, plus removal costs. This might have seemed a reasonable compensation cap at the time. However, despite the fact that the exact scale of damage caused by the Deepwater Horizon drilling rig explosion on 20 April 2010 remains unknown, it is clear that the costs of private economic and public natural-resource claims far exceed the $75-million cap on existing oil-spill legislation. Consider that even the costs for damaging natural resources and for private parties' claims in the much smaller case of the Exxon Valdez reached $2.3 billion. So what is required is to reconsider the design of such legislation, and this is why, after the Deepwater Horizon disaster, the White House, correctly, sought to raise the cap, while BP pledged to waive it, perhaps in an attempt to avoid the highest payments due to violations of safety regulations.[3]

Of course, by coping with the positive and negative risks of technologies, both legislation and safety technologies may still run into positive and negative risks of their own. But there is no problem of a *regressus ad infinitum* here, since handling metatechnological risks is no longer a technological issue, but an ethical one. What to privilege, how to find and allocate limited resources, which risks run by whom might be deemed acceptable by whom in view of whose advantages: these and similar questions do not have uncontroversial answers and they are not technological. They are open problems that require informed, reasonable, and tolerant debate, and an open mind. A philosophical attitude, in other words.

Clearly, there are no risk-free technologies, not even in an Amish-style approach to life, because technologies push the limit of the feasible and this, inevitably, comes at some risk. The only technologies completely safe are those never built. And there are no cost-free solutions for the management of technological risks either. But it is equally clear that there are metatechnological ways of dealing successfully with the risks implicit in any technology. And this is where smart ICTs become essential. We should invest further and more wisely in our metatechnologies: education, as the 'technology' that can improve people's minds, as we saw in Chapter 3; legislation, as the 'technology' that can improve social interactions; and of course smart second- and third-order meta-ICTs, which regulate and monitor other technologies. We need all this because the future will only be more technologically complex and challenging than the past.

There remains of course a significant risk, which is equally time-related. By developing smart ICTs that can help us to cope with the anthropocene's environmental costs we are taking a gambit, as I shall explain in the next section.

The gambit

Smart ICTs may play a huge role in our current environmental crisis, but there is a significant glitch. They can also be highly energy-consuming and hence potentially unfriendly towards the environment. As usual, some specific data will help to clarify the point.

In 2012, data warehouses around the world consumed about 30 billion watts of electricity every year. This is roughly equivalent to the output of 30 nuclear power plants. To have a sense of the magnitude, a single data centre can consume more energy than a medium-size town.[4] Things have been escalating in recent years, in connection with the huge amount of data we have been producing. Zettabytes need zettawatts. In 2000, data centres consumed 0.6 per cent of the world's electricity. In 2005, the figure had risen to 1 per cent. In 2007, ICTs accounted for the emission of 830 million tonnes of carbon dioxide, more or less 2 per cent of the estimated total, and approximately the same as the aviation industry's contribution. It is estimated that ICT-related emissions will increase by about 6 per cent a year until 2020, when they will greatly surpass the aviation's carbon footprint. ICTs are now responsible for more carbon-dioxide emissions per year than Argentina or the Netherlands.[5] Unsurprisingly, given this scenario, many data centres in California are listed in the state government's Toxic Air Contaminant Inventory, as major polluters.[6] In 2013, Google bought Sweden's entire wind-farm energy output (72 MW) in order to power its Finnish data centre and seek to remain carbon-neutral.[7]

It is not a green picture, and yet, we also know that ICTs can be a major ally in the environmental battle. McKinsey estimated that ICTs will help

> to eliminate 7.8 metric gigatons of greenhouse gas emissions annually by 2020 equivalent to 15% of global emissions today and five times more than our estimate of the emissions from these technologies in 2020.[8]

This is a positive but also much improvable balance. Still according to McKinsey, between 2008 and 2012 there has been little improvement in the industry's efficient use of energy, which is a mere 6 per cent to 12 per cent of the total power consumed.

The pros and contras of ICTs' environmental effects were well clarified by a report published by The Climate Group in 2008, entitled *SMART 2020: Enabling the Low Carbon Economy in the Information Age*:

> The ICT sector's own emissions are expected to increase, in a business as usual (BAU) scenario, from 0.53 billion tonnes (Gt) carbon dioxide equivalent (CO2e) in 2002 to 1.43 GtCO2e in 2020. But specific ICT opportunities identified in this report can lead to emission reductions five times the size of the sector's own footprint, up to 7.8 GtCO2e, or 15% of total BAU emissions by 2020.[9]

It is easier to see what all this means by looking at Figure 22.

The overall result is that we are taking a technological gambit: we are counting on the fact that ICTs benefit the environment more *significantly* and *quickly* than they actually harm it, and that there is enough time for such a gambit to pay back. The time variable is crucial, as I shall argue presently.

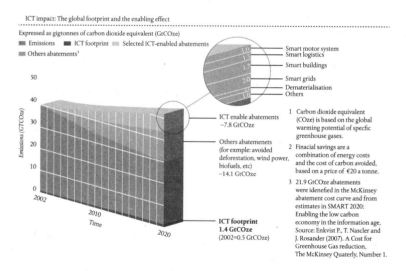

Fig. 22. ICTs and Their Environmental Impact.
Reproduced with permission, courtesy of The Climate Group 2009.

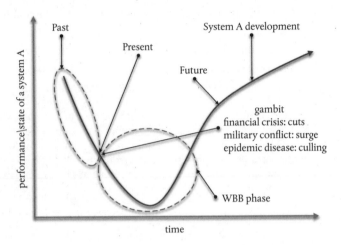

Fig. 23. The Logic of Worse Before Better (WBB).

A gambit is a chess opening in which a minor piece, often a pawn, is sacrificed to gain an advantage. It is therefore a matter of voluntary risk, involving a significant loss, taken strategically, in order to gain a significant advantage, that is higher than, and compensates for, the original loss. Such features characterize more generally the logic of 'worse before better'. Cuts during a financial crisis, a surge during a military conflict, culling to fight an epidemic disease, or chemotherapy to treat cancer[10] are all cases of implementation of such a logic. Figure 23 illustrates the general pattern.

In the case of ICTs, the gambit consists in causing some carbon emissions (losing the pawn) while trying to lower the global carbon footprint (winning the game). Of course, this is also a sound economic strategy for any company running data centres. It remains a gambit because we are betting that there is enough time to improve our ICTs so that they will raise the environmental well-being, thus enabling us to reap the benefits of such a strategy (see Figure 24).

If taken intelligently, the gambit should pay back. But for the gambit to be a success, we must make sure it is not a mere mistake (we lost

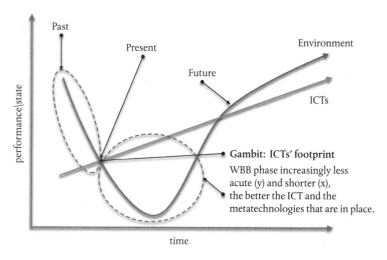

Fig. 24. ICTs and the Green Gambit.

the pawn accidentally or through miscalculation) but a successful strategy, and that the strategy takes into account the time variable, so that there is enough time to recoup the incurred losses. This means investing in green computing now. We urgently need more ICTs that are more environmentally sustainable (lower footprint) and that sustain the environment better (higher friendliness), especially as metatechnologies. They should help us to do 'more with less' (better use of available resources), 'more with left' (recycling of otherwise wasted resources), 'more with different' (using alternative resources, for instance through the dematerialization of supports). And all this more safely, not only by themselves but, above all, as metatechnologies that can regulate other technologies to achieve the indispensable levels of efficiency and safety required to benefit from the gambit and win the game. Nobody should have any illusion about the risky nature of the gambit itself. But better understanding and management of its risky nature and potential payoffs should make it a smarter move.

Conclusion

The greenest machine is a machine with 100 per cent energy efficiency. Unfortunately, this is equivalent to a perpetual motion machine and the latter is simply a pipe dream. This is not to say that such an impossible limit cannot be increasingly approximated. Energy waste can be dramatically reduced and energy efficiency can be highly increased (the two processes are not necessarily the same; compare recycling vs. doing more with less). Often, both kinds of processes may be fostered by relying on significant improvements in the management of information, especially by building and running hardware and processes better. ICTs can help us in our fight against the destruction, impoverishment, vandalism, and waste of both natural and human resources, including historical and cultural ones. So they can be a precious ally in what I have called *synthetic environmentalism* or *e-nvironmentalism*.[11] So this is how we may reinterpret Socrates' ethical intellectualism: we do evil because we do not know better, in the sense that the better our information management is, the less moral evil we may cause. The challenge is to reconcile our roles as agents within nature and as stewards of nature. We must develop into the right kind of demiurges. The good news is that it is a challenge we can meet.

The energy problems we are currently facing are not going to disappear. If anything, they are being exacerbated by the industrialization of an increasing number of countries, by the rising living standards of their populations, by ever more pressing issues related to global warming, and by the growth of the infosphere. We should address them now and decisively, from a metatechnological perspective, if we wish to tackle them before they become unmanageable or, even irreversible. And we should probably be ready to make sacrifices in terms of consumption and costs, if our ethical analyses of current and foreseeable metatechnological risks demand them. A better world might well be a more demanding one, both morally and economically. 'Pain now' may turn out to be the only successful strategy left.

10

ETHICS

E-nvironmentalism

From the Sumerian to the iPod's wheel, one way of recounting the story of technology is in terms of the increasingly faster evolution of layers of first-, second-, and third-order systems, vertically inter-dependent and horizontally integrated, relentlessly replacing, refining, complementing, and interacting with each other. If there is little trace of a philosophy of technology in antiquity it is probably because of the presence of only first-order technologies. With the advent of second-order technologies, such a macroscopic growth of in-betweenness could not escape philosophical scrutiny. Thus, modern philosophy is also a mechanical philosophy, as in Hobbes; a philosophy of dynamic mechanisms, as in Hegel or Marx; or a criticism of a mechanical culture, as in Heidegger, Foucault, or Lyotard. Contemporary philosophy, understood as the counterpart of third-order technologies, is still in the making, but I bet (a safe bet, since I will not be here be proved wrong—a trick I learnt from the best futurologists) that our future historian, whom we encountered in Chapter 1, will write on the pervasive presence in our time of an informational way of thinking: from network-oriented approaches to understanding society to input-elaboration-output schemes of processes, from distributed forms of agency to views of relations as dynamic interactions, from interface-like ways of understanding knowledge to data-based and software-driven ideas of science, and so forth. She will be struck by our influential new ways of conceptualizing ourselves, our world, and our culture hyperhistorically and informationally, no longer historically and mechanically.

For some time, the frontier of cyberspace has been the human/machine interface. For this reason, we have often regarded ourselves as lying outside of it. You will recall that, in his famous test, Turing posited a keyboard/screen interface to blanket human and computer. Half a century later, that interface has become part of our everyday reality. Helped perhaps by the ubiquitous television and the role it has played in informing and entertaining us, we now rely on interfaces as our second skins for communication, information, business, entertainment, socialization, and so forth. We have moved inside the infosphere. Its all-pervading nature also depends on the extent to which we accept its interface as integral to our reality and transparent to us, in the sense of no longer perceived as present. What matters is not so much moving bits instead of atoms—this is an outdated, communication-based interpretation of the information society that owes too much to mass-media sociology—as the far more radical fact that our understanding and conceptualization of the essence and fabric of reality is changing. Indeed, we have begun to accept the virtual as partly real and the real as partly virtual. As we saw in Chapter 3, the information society is better seen as a neo-manufacturing society in which raw materials and energy have been superseded by data and information, the new digital gold and the real source of added value. Not just communication and transactions then, but the creation, design, and management of information are the keys to the proper understanding of our hyperhistorical predicament and to the development of a sustainable infosphere.

Such understanding requires a new narrative, that is, a new sort of story we tell ourselves about our predicament and the human project we wish to pursue. This may seem an anachronistic step in the wrong direction. Until recently, there was much criticism of 'big narratives', from Marxism and liberalism to the so-called 'end of history'. But the truth is that such a criticism, too, was just another narrative, and it did not work. A systematic critique of grand narratives is inevitably part of the problem it tries to solve. Understanding why there are narratives, what justifies them, and what better narratives may replace them, is less juvenile and a more fruitful way ahead.

ICTs are creating the new informational environment in which future generations will live most of their time. Previous revolutions in the creation of wealth, especially the agricultural and the industrial ones, led to macroscopic transformations in our social and political structures and architectural environments, often without much foresight, normally with deep conceptual and ethical implications. The information revolution—whether understood as a third one, in terms of wealth creation, or as a fourth one, in terms of a reconceptualization of ourselves—is no less dramatic. We shall be in serious trouble, if we do not take seriously the fact that we are constructing the new physical and intellectual environments that will be inhabited by future generations. In view of this important change in the sort of ICT-mediated interactions that we will increasingly enjoy with other agents, whether biological or artificial, and in our self-understanding, an environmental approach seems a fruitful way of tackling the new ethical challenges posed by ICTs.[1] It is an approach that does not privilege the natural or untouched, but treats as authentic and genuine all forms of existence and behaviour, even those based on artificial, synthetic, hybrid, and engineered artefacts. The task is to formulate an ethical framework that can treat the infosphere as a new environment worthy of the moral attention and care of the human inforgs inhabiting it. Such an ethical framework must address and solve the unprecedented challenges arising in the new environment. It must be an *e-nvironmental ethics* for the whole infosphere. This sort of *synthetic* (both in the sense of holistic or inclusive, and in the sense of artificial) *environmentalism* will require a change in how we perceive ourselves and our roles with respect to reality, what we consider worth our respect and care, and how we might negotiate a new alliance between the natural and the artificial. It will require a serious reflection on the human project and a critical review of our current narratives, at the individual, social, and political levels. These are all pressing issues that deserve our full and undivided attention. Unfortunately, I suspect it will take some time and a whole new kind of education and sensitivity to realize that the infosphere is a common space, which needs to be preserved to the advantage of all. My hope is that this book may contribute to such a change in perspective.

FURTHER READING

Preface

If you wish to know more about the philosophers mentioned in this book, you may consider reading Magee (2000), very accessible. Floridi (2011) and Floridi (2013) are graduate-level treatments of the foundations of the philosophy of information and of information ethics.

Chapter 1

On writing as a technology and on the interplay between orality and literacy, the now classic reference is Ong (1988). Claude Shannon (1916–2001) is the father of information theory. His seminal work, Shannon and Weaver (1949, rep. 1998), requires a good background in mathematics and probability theory. An accessible introduction to information theory is still Pierce (1980). I have covered topics discussed in this chapter in Floridi (2010a), where the reader can also find a simple introduction to information theory. Gleick (2011) is an enjoyable 'story' of information. Ceruzzi (2012) provides a short introduction to the history of computing, from its beginning to the Internet. Caldarelli and Catanzaro (2012) give a short introduction to networks. On big data, a good survey is O'Reilly Media (2012), the Kindle edition is free. Mayer-Schönberger and Cukier (2013) is accessible. On the postmodern society, Lyotard (1984) is essential reading, philosophically demanding but also rewarding. On the network society, Manuel Castells (2000), the first volume of his trilogy, has shaped the debate. The information society produces much information about itself. Among the many valuable, yearly reports, freely available online, one may consult *Measuring the Information Society*, which includes the ICT Development Index and the ICT Price Basket, two benchmarks useful for monitoring the development of the information society worldwide; the *Global Information Technology Report*, produced by the World Economic Forum in cooperation with INSEAD, which covers 134 economies worldwide and is acknowledged to be the most comprehensive and authoritative international assessment of the impact of ICT on countries' development and competitiveness; the *International Telecommunication Union Statistics*, which collects data about connectivity and availability of telecommunication services worldwide; and the *Digital Planet* report, published by World

Information Technology and Service Alliance, which contains projections on ICT spending. Finally, Brynjolfsson and McAfee (2011) analyse how ICTs affect the job market, transform skills, and reshape the evolution of human labour. They do so from an American perspective, but their insights are universal, the scholarship admirable, and the recommendations on how machines and humans may collaborate quite convincing.

Chapter 2

A useful, short, but comprehensive overview of the history of technology is offered by Headrick (2009). For a textbook-like approach, oriented towards the interactions between technology and science, a good starting point is McClellan and Dorn (2006). Shumaker et al. (2011) provide an important reference in the animal tool-making behaviour literature; the first edition published in 1980 was influential. A first and easy introduction to the history of inventions is offered by Challoner (2009), part of the 1001 series. The history of interfaces is ripe for a good introductory book, as the ones that I know are all technical. On the visualization of information, a classic is Tufte (2001), which could be accompanied by McCandless (2012). Design is another immense field of studies. As a starting point, one may still choose Norman (1998) although it is now slightly outdated (it is basically a renamed version of *The Psychology of Everyday Things*, published in 1989 by the same author). Millenials are described in Howe and Strauss (2000), but see also Palfrey and Gasser (2008) on digital natives. On globalization, I would recommend another title in the Very Short Introductions series, by Steger (2003). Some of the most important ideas—including that of authenticity—concerning the influence of mechanical reproduction of objects on our understanding and appreciation of their value are discussed in *The Work of Art in the Age of Mechanical Reproduction*, the classic and influential work by Walter Benjamin; for a recent translation see W. Benjamin (2008).

Chapter 3

This chapter is loosely based on chapter 11 in Floridi (2013). On the philosophy of personal identity, a rigorous and accessible introduction is Noonan (2003). A simpler overview of the philosophy of mind is offered by Feser (2006). On multi-agent system, a great book is Wooldridge (2009). Weinberger (2011) is a valuable book on how ICTs are changing knowledge and its processes. I would also strongly recommend on similar themes Brown and Duguid (2002). Floridi (2014) is a collection of essays addressing the onlife experience. Sainsbury (1995) is the standard reference for a scholarly treatment of paradoxes. If you wish to read something lighter and more entertaining Smullyan (1980) is still a good choice.

Chapter 4

This chapter in loosely based on chapter 1 in Floridi (2013). The idea of the first three revolutions is introduced by Freud (1917), for a scholarly analysis see Weinert (2009). On Alan Turing, an excellent introduction is still Hodges (1992), which is essentially a reprint of the 1983 edition. On Turing's influence, Bolter (1984) is a gem that has been unfairly forgotten. The view that we may be becoming cyborgs is articulated by Clark (2003). Davis (2000) is a clear and accessible overview of the logical, mathematical, and computational ideas that led from Leibniz to Turing. Goldstine (1993) has become almost a classic reference for the history of the computer, first published in 1972; the fifth paperback printing (1993) contains a new preface.

Chapter 5

This chapter in loosely based on chapter 12 in Floridi (2013). Wacks (2010) offers a short introduction to privacy. For a more philosophical treatment, including the analysis of privacy in public spaces, see Nissenbaum (2010). A sophisticated analysis of the networked self is offered by Cohen (2012). For a lively discussion of security issues and how to balance them with civil rights see Schneier (2003).

Chapter 6

Turing (2004) is a collection of his most important writings. It is not for the beginner, who may wish to start by reading Copeland (2012). Shieber (2004) contains an excellent collection of essays on the Turing Test. Negnevitsky (2011) is a simple and accessible introduction to artificial intelligence, lengthy but also modular. Norbert Wiener (1894–1964) was the father of cybernetics. He wrote extensively and insightfully on the relations between humanity and its new machines. His three works, Wiener (1954, 1961, 1964), are classics not to be missed. Winfield (2012) is a short introduction to robotics. A graduate-level discussion of the symbol grounding problem can be found in Floridi (2011). Two great and influential works that discuss the nature of artificial intelligence from different perspectives are Weizenbaum (1976), the designer of ELIZA, and Simon (1996). A slightly dated but still useful criticism of strong AI is provided by Dreyfus (1992).

Chapter 7

Han (2011) is an accessible text on Web 2.0, Antoniou (2012) is a more demanding introduction to the Semantic Web. Dijck (2013) offers a critical reconstruction of social media. For a detailed critique of the Semantic Web see Floridi (2009).

Chapter 8

This chapter is loosely based on Floridi (2014). Linklater (1998) gives an account of post-Westphalian societies and the ethical challenges ahead. On Bretton Woods and the emergence of our contemporary financial and monetary system, see Steil (2013). Clarke and Knake (2010) approach the problems of cyberwar and cyber security from a political perspective that would still qualify as 'historical' within this book, but it is helpful. Floridi and Taddeo (2014) is a collection of essays exploring the ethics of cyberwar. Floridi (2013) offers a foundational analysis of information ethics. An undergraduate-level introduction to problems and theories in information and computer ethics is Floridi (2010b). On politics and the information society, two recommendable readings are Mueller (2010) and Brown and Marsden (2013). The idea that there are four major regulators of human behaviour—law, norms, market, and architecture—was influentially developed by Lessig (1999), see also Lessig (2006).

Chapter 9

To understand the nature and logic of risk, a good starting point is the short introduction by Fischhoff and Kadvany (2011). Although Hird (2010) is not aimed at the educated public but more to CEOs of big companies, he offers a good overview of green computing, its problems and advantages. N. Carter (2007) guides the reader through environmental philosophy and green political thinking, environmental parties and movements, as well as policymaking and environmental issues.

Chapter 10

The following four books are a bit demanding but deserve careful reading, if you wish to understand the problems discussed in this book in more depth: Wiener (1954), Weizenbaum (1976), Lyotard (1984), Simon (1996). They belong to different intellectual traditions. Each of them has profoundly influenced your author.

REFERENCES

Aldridge, I. (2013). *High-frequency trading: A practical guide to algorithmic strategies and trading systems* (2nd edn.). Hoboken, NJ: Wiley.

Anderson, C. (23 June 2008). The end of theory: Data deluge makes the scientific method obsolete. *Wired Magazine*.

Anderson, R., and Moore, T. (2006). The economics of information security. *Science*, 314 (5799), 610–13.

Anderson, S., and Cavanagh, J. (2000). Top 200: The rise of corporate global power. *Institute for Policy Studies*, 4.

Antoniou, G. (2012). *A Semantic Web primer* (3rd edn.). Cambridge, Mass. and London: MIT Press.

Aristotle. (1995). *The politics*. Oxford and New York: Oxford University Press.

Asimov, I. (1956). The dead past. *Astounding Science Fiction* (April), 6–46.

Ata, R. N., Thompson, J. K., and Small, B. J. (2013). Effects of exposure to thin-ideal media images on body dissatisfaction: Testing the inclusion of a disclaimer versus warning label. *Body image*, 10(4), 472–80.

Barboza, D. (9 December 2005). Ogre to slay? Outsource it to Chinese. *New York Times*, A1.

Barnatt, C. (2010). *A brief guide to cloud computing: An essential introduction to the next revolution in computing*. London: Robinson.

BBC News. (11 October 2007). Speaker's legal costs criticised.

BBC News. (7 June 2013). Self-portraits and social media: The rise of the 'selfie'.

BBC News. (9 August 2013). The pirate bay: BitTorrent site sails to its 10th birthday.

BBC News. (16 September 2013). Netflix studies piracy sites to decide what to buy.

BBC News. (2 December 2013). Amazon testing drones for deliveries.

Benjamin, M. (2013). *Drone warfare: Killing by remote control* (updated edn.). London and New York: Verso.

Benjamin, W. (2008). *The work of art in the age of mechanical reproduction*. London: Penguin.

Bentham, J. (2011). *Selected writings*. New Haven: Yale University Press.

Berkman, M. B., and Plutzer, E. (2010). *Evolution, creationism, and the battle to control America's classrooms*. New York: Cambridge University Press.

Berners-Lee, T., Hendler, J., and Lassila, O. (2001). The Semantic Web. *Scientific American*, 284(5), 28–37.

Boccaletti, G., Löffler, M., and Oppenheim, J. M. (2008). How IT can cut carbon emissions. *McKinsey Quarterly* (October), 1–5.

Bolter, J. D. (1984). *Turing's man: Western culture in the computer age*. London: Duckworth.

Bond, M., Meacham, T., Bhunnoo, R., and Benton, T. G. (2013). Food waste within global food systems. *A Global Food Security Report*, available online.

Brezis, E. S., Krugman, P. R., and Tsiddon, D. (1993). Leapfrogging in international competition: A theory of cycles in national technological leadership. *American Economic Review*, 83(5), 1211–19.

Bridle, J. (29 September 2013). Matchbook: One giant leap towards a virtual book collection. *The Observer*.

Briere, M., Oosterlinck, K., and Szafarz, A. (2013). Virtual currency, tangible return: Portfolio diversification with bitcoins. *Working Papers CEB*, 13 (Université Libre de Bruxelles).

Brown, I., and Marsden, C. T. (2013). *Regulating code: Good governance and better regulation in the information age*. Cambridge, Mass.: MIT Press.

Brown, J. S., and Duguid, P. (2002). *The social life of information* (repr. with a new preface, originally 2000 edn.). Boston: Harvard Business School; McGraw-Hill.

Brynjolfsson, E., and McAfee, A. (2011). *Race against the machine: How the digital revolution is accelerating innovation, driving productivity, and irreversibly transforming employment and the economy*. Lexington, Mass.: Digital Frontier Press.

Caldarelli, G., and Catanzaro, M. (2012). *Networks: A very short introduction*. Oxford: Oxford University Press.

Carr, N. (1 July 2008). Is Google making us stupid? *The Atlantic*.

Carter, C. (12 September 2013). Just 224 tweets for modern-day couples to fall in love. *The Telegraph*.

Carter, N. (2007). *The politics of the environment: Ideas, activism, policy* (2nd edn.). Cambridge: Cambridge University Press.

Castells, M. (2000). *The rise of the network society* (2nd edn.). Oxford: Blackwell.

Ceruzzi, P. E. (2012). *Computing: A concise history*. Cambridge, Mass. and London: MIT Press.

Challoner, J. (2009). *1001 inventions that changed the world*. Hauppauge, NY: Barron's.

Clark, A. (2003). *Natural-born cyborgs: Minds, technologies, and the future of human intelligence*. Oxford and New York: Oxford University Press.

Clarke, R. A., and Knake, R. K. (2010). *Cyber war: The next threat to national security and what to do about it*. New York: Ecco.

Clemens, A. M. (2004). No computer exception to the constitution: The Fifth Amendment protects against compelled production of an encrypted document or private key. *UCLA Journal of Law and Technology*, 2–5.

Cohan, P. (20 May 2013). Yahoo's Tumblr buy fails four tests of a successful acquisition. *Forbes*.

Cohen, J. (2000). Examined lives: Informational privacy and the subject as object. *Stanford Law Review*, 52: 1373–1437.

Cohen, J. (2012). *Configuring the networked self: Law, code, and the play of everyday practice*. New Haven: Yale University Press.

Copeland, B. J. (2012). *Turing: Pioneer of the information age*. Oxford: Oxford University Press.

Cosmides, L. (1989). The logic of social exchange: Has natural selection shaped how humans reason? Studies with the Wason Selection Task. *Cognition*, 31(3), 187–276.

Crutzen, P. J. (2006). The 'anthropocene'. In E. Ehlers and T. Krafft (eds.), *Earth system science in the anthropocene* (pp. 13–18): Heidelberg: Springer.

Cukier, K. (27 February 2010). Data, data everywhere: A special report on managing information: *The Economist*, 3–18.

Dannenberg, R. A. (ed.). (2010). *Computer games and virtual worlds: A new frontier in intellectual property law*. Chicago: American Bar Association, Section of Intellectual Property Law.

Davis, M. (2000). *Engines of logic: Mathematicians and the origin of the computer*. New York and London: Norton.

Dibbell, J. (17 June 2007). The life of the Chinese gold farmer. *New York Times*.

Dijck, J. v. (2013). *The culture of connectivity: A critical history of social media*. Oxford: Oxford University Press.

Dijkstra, E. W. (1984). The threats to computing science. Paper presented at the ACM 1984 South Central Regional Conference, 16–18 November, Austin, Tex.

Doward, J., and Hinsliff, G. (24 February 2008). Air miles, taxis and his reluctance to come clean bring Speaker to the brink of disgrace. *The Observer*.

Dredge, S. (27 June 2013). Autorip comes to the UK: Amazon's 'gentle bridge' between real and virtual music. *The Guardian*.

Dreyfus, H. L. (1992). *What computers still can't do: A critique of artificial reason* (rev. edn.). Cambridge, Mass.: MIT Press.

The Economist. (16 December 1999). Living in the global goldfish bowl.

The Economist. (20 December 2005). Frequent-flyer miles—funny money.

The Economist. (7 June 2007). Robot wars.

The Economist. (22 May 2008). Down on the server farm.

The Economist. (29 February 2012). Now for the good news.

The Economist. (2 June 2012). Morals and the machine.

Evans, D. (2011). The Internet of things: How the next evolution of the Internet is changing everything. *CISCO white paper* (April).

Feser, E. (2006). *Philosophy of mind: A beginner's guide*. Oxford: Oneworld.

Fiddick, L., Cosmides, L., and Tooby, J. (2000). No interpretation without representation: The role of domain-specific representations and inferences in the Wason Selection Task. *Cognition*, 77(1), 1–79.

Fineman, M., and Mykitiuk, R. (1994). *The public nature of private violence: The discovery of domestic abuse.* New York and London: Routledge.

Fischhoff, B., and Kadvany, J. D. (2011). *Risk: A very short introduction.* Oxford: Oxford University Press.

Fishman, T. C. (2010). *Shock of gray: The aging of the world's population and how it pits young against old, child against parent, worker against boss, company against rival, and nation against nation.* New York: Scribner.

Floridi, L. (1999). *Philosophy and computing: An introduction.* London and New York: Routledge.

Floridi, L. (2009). The Semantic Web vs. Web 2.0: A philosophical assessment. *Episteme,* 6, 25–37.

Floridi, L. (2010a). *Information: A very short introduction.* Oxford: Oxford University Press.

Floridi, L. (ed.). (2010b). *The Cambridge handbook of information and computer ethics.* Cambridge: Cambridge University Press.

Floridi, L. (2011). *The philosophy of information.* Oxford: Oxford University Press.

Floridi, L. (2012). Acta—the ethical analysis of a failure, and its lessons. *ECIPE working papers,* 04/2012.

Floridi, L. (2013). *The ethics of information.* Oxford: Oxford University Press.

Floridi, L. (ed.). (2014). *The onlife manifesto.* New York: Springer.

Floridi, L., and Taddeo, M. (eds.). (2014). *The ethics of information warfare.* New York: Springer.

Fowler, W., North, A., and Stronge, C. (2008). *The history of pistols, revolvers and submachine guns: The development of small firearms, from 12th century hand-cannons to modern-day automatics, with 180 fabulous photographs and illustrations.* London: Southwater.

Freud, S. (1917). A difficulty in the path of psycho-analysis. *The Standard Edition of the Complete Psychological Works of Sigmund Freud,* ed. James Strachey (London: Hogarth Press, 1956–74), xvii. *An Infantile Neurosis and Other Works, 1917–1919,* 135–44.

Fuggetta, R. (2012). *Brand advocates: Turning enthusiastic customers into a powerful marketing force.* Hoboken, NJ: Wiley.

Funabashi, H. (2012). Why the Fukushima nuclear disaster is a man-made calamity. *International Journal of Japanese Sociology,* 21(1), 65–75.

Gabbatt, A., and Rushe, D. (3 October 2013). Silk Road shutdown: How can the FBI seize Bitcoins? *The Guardian.*

Gantz, J., and Reinsel, D. (2011). *Extracting value from chaos.* White paper sponsored by EMC–IDC, available online.

Gardiner, B. (15 August 2008). Bank failure in second life leads to call for regulation. *Wired Magazine.*

Gardner, S., and Krug, K. (2006). *BitTorrent for dummies.* Hoboken, NJ: Wiley Pub.

Glanz, J. (22 September 2012). Power, pollution and the Internet. *New York Times.*

Gleick, J. (2011). *The information: A history, a theory, a flood.* London: Fourth Estate.

Gogol, N. (1998). *Dead souls*. Oxford: Oxford University Press.

Goldstine, H. H. (1993). *The computer from Pascal to von Neumann*. Princeton: Princeton University Press.

Halper, S. A. (2010). *The Beijing Consensus: How China's authoritarian model will dominate the twenty-first century*. New York: Basic Books.

Han, S. (2011). *Web 2.0*. London: Routledge.

Harrop, P., and Das, R. (2013). RFID forecasts, players and opportunities 2012–2022. *IDTechEx, Cambridge, UK*.

Headrick, D. R. (2009). *Technology: A world history*. Oxford: Oxford University Press.

Hegel, G. W. F. (1979). *Phenomenology of spirit*. Oxford: Oxford University Press.

Heilman, J. M., Kemmann, E., Bonert, M., Chatterjee, A., Ragar, B., Beards, G. M., et al. (2011). Wikipedia: A key tool for global public health promotion. *Journal of Medical Internet Research*, 13(1).

Herzfeld, O. (4 December 2012). What is the legal status of virtual goods? *Forbes*.

Hill, K. (4 July 2012). How Target figured out a teen girl was pregnant before her father did. *Forbes*.

Hird, G. (2010). *Green IT in practice: How one company is approaching the greening of its IT* (2nd edn.). Ely: IT Governance Publishing.

Hobbes, T. (1991). *Leviathan*. Cambridge: Cambridge University Press.

Hockenos, P. (2011). *State of the world's volunteerism report, 2011: Universal values for global well-being*. New York: United Nations, available online.

Hodges, A. (1992). *Alan Turing: The Enigma*. London: Vintage.

Holvast, J. (2009). History of privacy. In V. Matyáš, S. Fischer-Hübner, D. Cvrček, and P. Švenda (eds.), *The future of identity in the information society (IFIP advances in information and communication technology, 298)*. Berlin and Heidelberg: Springer, pp. 13–42.

Honan, M. (5 August 2013). Remembering the Apple Newton's prophetic failure and lasting impact. *Wired Magazine*.

Horace (2011). *Satires and epistles*. New York: Oxford University Press.

Horowitz, S. J. (2008). Bragg v. Linden's second life: A primer in virtual world justice. *Ohio NUL Review*, 34, 223.

Howe, N., and Strauss, W. (2000). *Millennials rising: The next great generation*. New York: Vintage Books.

James, W. (1890). *The principles of psychology*. London: Macmillan.

Juniper Research. (2012). *Smart wearable devices*. Research report. Hampshire, UK.

Klein, N. (2000). *No logo: No space, no choice, no jobs*. London: Flamingo.

Lenhart, A. (19 March 2012). Teens, smartphones & texting. *Pew Internet & American Life Project*.

Lenhart, A., and Madden, M. (18 April 2007). Teens, privacy & online social networks: How teens manage their online identities and personal information in the age of myspace: *Pew Internet & American Life Project*.

Lessig, L. (1999). *Code: And other laws of cyberspace*. New York: Basic Books.

Lessig, L. (2006). *Code* (2nd edn.). New York: BasicBooks.

Linklater, A. (1998). *The transformation of political community: Ethical foundations of the post-Westphalian era.* Oxford: Polity.

Logan, H. C. (1944). *Hand cannon to automatic: a pictorial parade of hand arms.* Huntington, W.Va.: Standard Publications, Incorporated.

Lohr, S. (10 March 2013). Algorithms get a human hand in steering web. *New York Times.*

Luke, K. (2013). World of warcraft down to 7.7 million subscribers. *IGN,* available online.

Lyman, P., and Varian, H. R. (2003). *How much information 2003,* available online.

Lyotard, J.-F. (1984). *The postmodern condition: A report on knowledge.* Minneapolis: University of Minnesota Press.

McCandless, D. (2012). *Information is beautiful.* London: Collins.

McCarthy, J. (1997). AI as sport. *Science and Engineering Ethics,* 276(5318), 1518–19.

McClellan, J. E., and Dorn, H. (2006). *Science and technology in world history: An introduction* (2nd edn.). Baltimore: Johns Hopkins University Press.

Magee, B. (2000). *The great philosophers: An introduction to Western philosophy* (2nd edn.). Oxford: Oxford University Press.

Martin, L. H. (ed.). (1988). *Technologies of the self: A seminar with Michel Foucault.* Amherst, Mass.: University of Massachusetts Press.

Marwick, A., Murgia-Diaz, D., and Palfrey, J. (2010). Youth, privacy and reputation (literature review). *Berkman Center Research Publication* 2010-5; *Harvard Law Working Paper* 10-29.

Mason, M. (6 May 2013). Reports of our death have been greatly exaggerated. *BitTorrent Blog.*

Mayer-Schönberger, V., and Cukier, K. (2013). *Big data: A revolution that will transform how we live, work, and think.* Boston: Houghton Mifflin Harcourt.

Mills, E. (2005). Google balances privacy, reach. *C|Net News.com,* <http://news.com.com/Google+balances+privacy%2C+reach/2100-1032_3-5787483.html>.

Mueller, M. (2010). *Networks and states: The global politics of internet governance.* Cambridge, Mass.: MIT Press.

Muniz Jr, A. M., and Schau, H. J. (2005). Religiosity in the abandoned Apple Newton brand community. *Journal of Consumer Research,* 31(4), 737–47.

NATO Cooperative Cyber Defence Centre of Excellence. (2013). *Tallinn manual on the international law applicable to cyber warfare: Prepared by the international group of experts at the invitation of the NATO Cooperative Cyber Defence Centre of Excellence.* Cambridge: Cambridge University Press.

Negnevitsky, M. (2011). *Artificial intelligence: A guide to intelligent systems* (3rd edn.). Harlow: Addison Wesley/Pearson.

Neurath, O. (1959). Protocol sentences. In A. J. Ayer (ed.), *Logical positivism* (pp. 199–208). Glencoe, Ill.: The Free Press.

Nissenbaum, H. F. (2010). *Privacy in context: Technology, policy, and the integrity of social life.* Stanford, Calif.: Stanford Law Books.

Nogee, A. (2004). RFID tags and chip: Changing the world for less than the price of a cup of coffee. *In-Stat/MDR*.

Noonan, H. W. (2003). *Personal identity* (2nd edn.). London: Routledge.

Norman, D. A. (1998). *The design of everyday things*. Cambridge, Mass.: MIT.

Norris, P. (2001). *Digital divide: Civic engagement, information poverty, and the Internet worldwide*. Cambridge: Cambridge University Press.

Noyes, J. (1983). The QWERTY keyboard: A review. *International Journal of Man-Machine Studies*, 18(3), 265–81.

O'Reilly Media. (2012). *Big data now: 2012 edition*. Kindle edition: O'Reilly Media Inc.

O'Reilly, T. (1 October 2005). Web 2.0: Compact definition? *O'Reilly radar blog*, available online.

Oldham, G. R., and Brass, D. J. (1979). Employee reactions to an open-plan office: A naturally occurring quasi-experiment. *Administrative Science Quarterly*, 24, 267–84.

Ong, W. J. (1988). *Orality and literacy: The technologizing of the word*. London: Routledge.

Orwell, G. (2013). *Nineteen Eighty-Four* (annotated edn.). London: Penguin.

Palfrey, J., and Gasser, U. (2008). *Born digital: Understanding the first generation of digital natives*. New York: Basic Books.

Pascal, B. (1997). Machine d'arithmétique. *Review of Modern Logic*, 7(1), 56–66.

Pascal, B. (2008). *Pensées and other writings*. Oxford: Oxford University Press.

Pfanner, E. (28 September 2009). A move to curb digitally altered photos in ads. *New York Times*.

Picard, R. W. (1997). *Affective computing*. Cambridge, Mass. and London: MIT Press.

Pierce, J. R. (1980). *An introduction to information theory: Symbols, signals & noise* (2nd edn.). New York: Dover Publications.

Plato. (1997). *Complete works*. Indianapolis and Cambridge, Mass.: Hackett.

Popper, K. R. (2002). *The logic of scientific discovery*. London: Routledge.

Proust, M. (1996). *In search of lost time*, i. *Swann's way* (trans. C. K. Scott Moncrieff and T. Kilmartin, rev. D. J. Enright). London: Vintage (originally published: London: Chatto & Windus, 1992).

Raice, S. (2 Feb. 2012). Facebook sets historic IPO. *Wall Street Journal*.

Ramo, J. C. (2004). *The Beijing Consensus*. London: Foreign Policy Centre.

Rawls, J. (1999). *A theory of justice* (rev. edn.). Cambridge, Mass.: Belknap Press of Harvard Univeristy Press.

Rose, S. (19 October 2009). Hollywood is haunted by ghost in the shell. *The Guardian*.

Sainsbury, R. M. (1995). *Paradoxes* (2nd edn.). Cambridge: Cambridge University Press.

Salinger, J. D. (1951). *The catcher in the rye*. Boston: Little Brown.

Saponas, T. S., Lester, J., Hartung, C., and Kohno, T. (2006). Devices that tell on you: The Nike+iPod sport kit. *Dept. of Computer Science and Engineering, University of Washington, Tech. Rep.,* available online.

Sauter, M. B., Hess, A. E. M., and Weigley, S. (1 February 2013). The biggest car recalls of all time. *24/7 Wall St. Report.*

Schaefer, M. (2012). *Return on influence: The revolutionary power of Klout, social scoring, and influence marketing* (1st edn.). New York: McGraw-Hill.

Schneier, B. (2003). *Beyond fear: Thinking sensibly about security in an uncertain world.* New York: Copernicus, an imprint of Springer.

Shannon, C. E., and Weaver, W. (1949, repr. 1998). *The mathematical theory of communication.* Urbana: University of Illinois Press.

Shieber, S. M. (2004). *The Turing Test: Verbal behavior as the hallmark of intelligence.* Cambridge, Mass.: MIT Press.

Shin, G.-W., and Sneider, D. C. (2011). *History textbooks and the wars in Asia: Divided memories.* Abingdon: Routledge.

Shumaker, R. W., Walkup, K. R., and Beck, B. B. (2011). *Animal tool behavior: The use and manufacture of tools by animals* (rev. and updated edn.). Baltimore: Johns Hopkins University Press.

Simon, H. A. (1996). *The sciences of the artificial* (3rd edn.). Cambridge, Mass.: MIT Press.

Slezak, M. (2013). Space miners hope to build first off-earth economy. *New Scientist,* 217(2906), 8–10.

Smullyan, R. M. (1980). *This book needs no title: A budget of living paradoxes.* Englewood Cliffs, NJ: Prentice-Hall.

Soltani, A., Canty, S., Mayo, Q., Thomas, L., and Hoofnagle, C. (2009). Flash cookies and privacy, available at SSRN 1446862.

Steger, M. B. (2003). *Globalization: A very short introduction.* Oxford: Oxford University Press.

Steil, B. (2013). *The battle of Bretton Woods: John Maynard Keynes, Harry Dexter White, and the making of a new world order.* Princeton: Princeton University Press.

Sterin, F. (2013). Powering our Finnish data center with Swedish wind energy. *Google Green Blog,* <http://googlegreenblog.blogspot.co.uk/2013/06/powering-our-finnish-data-center-with.html>.

Sundstrom, E., Herbert, R. K., and Brown, D. W. (1982). Privacy and communication in an open-plan office a case study. *Environment and Behavior,* 14(3), 379–92.

Turing, A. M. (1936). On computable numbers, with an application to the entscheidungsproblem. *Proceedings of the London Mathematics Society,* 2nd ser., 42, 230–65.

Turing, A. M. (1950). Computing machinery and intelligence. *Mind,* 59(236), 433–60.

Turing, A. M. (2004). *The essential Turing: Seminal writings in computing, logic, philosophy, artificial intelligence, and artificial life, plus the secrets of Enigma.* Oxford: Clarendon Press.

Uppenberg, K. (2009). Innovation and economic growth. *European Investment Bank Papers,* 14(1), 10–35.

US Civil Service. (1891). *The executive documents, house of representatives, first session of the Fifty-First Congress 1889–1890.* Washington, DC.

Van Duyn, A., and Waters, R. (7 August 2006). Google in $900m ad deal with myspace. *Financial Times.*

Vesset, D., Morris, H. D., Little, G., Borovick, L., Feldman, S., Eastwood, M., et al. (2012). *Worldwide big data technology and services 2012–2015 forecast,* a report by IDC, available online.

Von Ahn, L., Maurer, B., McMillen, C., Abraham, D., and Blum, M. (2008). Recaptcha: Human-based character recognition via web security measures. *Science,* 321(5895), 1465–8.

Wacks, R. (2010). *Privacy: A very short introduction.* Oxford: Oxford University Press.

Waismann, F. (1968). *How I see philosophy.* London: Macmillan (1st pub. 1956).

Warren, S., and Brandeis, L. D. (1890). The right to privacy. *Harvard Law Review,* 193(4).

Weaver, M. (13 May 2010). Obama to force BP to pay more cleanup costs for deepwater disaster. *The Guardian.*

Webb, M. (2008). Smart 2020: Enabling the low carbon economy in the information age. *The Climate Group,* available online.

Weinberger, D. (2011). *Too big to know: Rethinking knowledge now that the facts aren't the facts, experts are everywhere, and the smartest person in the room is the room.* New York: Basic Books.

Weinert, F. (2009). *Copernicus, Darwin, and Freud: Revolutions in the history and philosophy of science.* Oxford: Blackwell.

Weizenbaum, J. (1966). ELIZA—a computer program for the study of natural language communication between man and machine. *Communications of the ACM,* 9(1), 36–45.

Weizenbaum, J. (1976). *Computer power and human reason: From judgment to calculation.* San Francisco: W. H. Freeman.

Wheeler, J. A. (1990). Information, physics, quantum: The search for links. In W. H. Zurek (ed.), *Complexity, entropy, and the physics of information.* Redwood City, Calif.: Addison-Wesley Pub. Co., pp. 3–28.

Wiener, N. (1954). *The human use of human beings: Cybernetics and society* (rev. edn.). Boston: Houghton Mifflin.

Wiener, N. (1961). *Cybernetics: Or control and communication in the animal and the machine* (2nd edn.). Cambridge, Mass.: MIT Press.

Wiener, N. (1964). *God and golem, inc.: A comment on certain points where cybernetics impinges on religion.* Cambridge, Mass.: MIT Press.

Williamson, J. (1993). Democracy and the 'Washington Consensus'. *World Development*, 21(8), 1329–36.

Williamson, J. (2012). Is the 'Beijing Consensus' now dominant? *Asia Policy*, 13(1), 1–16.

Winfield, A. F. T. (2012). *Robotics: A very short introduction*. Oxford: Oxford University Press.

Wittgenstein, L. (2001). *Philosophical investigations: The German text with a revised English translation* (3rd edn.). Oxford: Blackwell.

Wolfe, C. K., and Akenson, J. E. (2005). *Country music goes to war*. Lexington, KY: University Press of Kentucky.

Wollslager, M. E. (2009). Children's awareness of online advertising on neopets: The effect of media literacy training on recall. *SIMILE: Studies In Media & Information Literacy Education*, 9(2), 31–53.

Wooldridge, M. J. (2009). *An introduction to multiagent systems* (2nd edn.). Chichester and Hoboken, NJ: Wiley.

Woolf, V. (1999). *The years*. Oxford: Oxford University Press.

Woolf, V. (2003). *The common reader* (rev. edn.). London: Vintage.

ENDNOTES

Preface

1. I have developed the project for a philosophy of information in Floridi (2011) and Floridi (2013).
2. Waismann (1968), p. 19.
3. Apparently, if you were in Vienna at that time and you disliked foundationalism, water was your friend. Karl Popper (1902–94), the great philosopher of science, born in Vienna, never was a member of the Vienna Circle, but had many contacts with it and, remarkably, used another aquatic metaphor to describe science: 'Science does not rest upon solid bedrock. The bold structure of its theories rises, as it were, above a swamp. It is like a building erected on piles. The piles are driven down from above into the swamp, but not down to any natural or "given" base; and if we stop driving the piles deeper, it is not because we have reached firm ground. We simply stop when we are satisfied that the piles are firm enough to carry the structure, at least for the time being.' Popper (2002), p. 94.
4. 'There is no way of taking conclusively established pure protocol sentences as the starting point of the sciences. No *tabula rasa* exists. We are like sailors who must rebuild their ship on the open sea, never able to dismantle it in dry-dock and to reconstruct it there out of the best materials. Only the metaphysical elements can be allowed to vanish without trace. Vague linguist conglomerations always remain in one way or another as components of the ship.' Neurath (1959), p. 201.

Acknowledgements

1. *The Manifesto* is available online. The final version, with commentaries and background chapters, is published in Floridi (2014).

Chapter 1

1. According to the French biologist Jean-Baptiste Lamarck (1744–1829), an organism could pass on to its offspring adaptive changes acquired through

individual efforts during its lifetime. This pre-Darwinian theory is known as soft inheritance.

2. Source: Survival for Tribal Peoples report, *Uncontacted Amazon Indians face annihilation*, 14 February 2011, available online.
3. Evans (2011), p. 3.
4. Lyman and Varian (2003).
5. One exabyte corresponds to 10^{18} bytes or a 50,000 year-long video of DVD quality.
6. Gantz and Reinsel (2011).
7. Source: NSF-12-499, available online.
8. Hill (4 July 2012).
9. This is the agency of the United States government that maintains government and historical records, including the legally authentic and authoritative copies of acts of Congress, presidential proclamations and executive orders, and federal regulations.
10. Another common measure that is becoming more popular is the Annualized Failure Rate (AFR), which indicates the estimated probability that a system will fail during a full year of use. It is a relation between the MTBF and the hours that a number of devices are run per year.
11. Source: IBIS World, *Data Recovery Services Market Research Report*, July 2012, available online.
12. For a short overview see Cukier (27 February 2010). For a more recent overview see Vesset et al. (2012) on which the report is based.
13. Source: ICT Data and Statistics Division, Telecommunication Development Bureau, International Telecommunication Union, *The World in 2013, ICT Facts and Figures*, available online.

Chapter 2

1. For some time, I thought that *affordance* would do, but this is a term that has other technical connotations in other specific contexts, so may prove too confusing. Using it would mean the sun would have to be described as an affordance *for* the hat, whereas we really want to say that blocking the sun is an affordance *of* the hat.
2. Brezis et al. (1993).
3. M. Benjamin (2013).
4. Aristotle (1995), 1.253b31–3.
5. Aristotle (1995), 1.1254a14–18.
6. Muniz Jr and Schau (2005), Honan (5 August 2013).
7. Aldridge (2013).
8. Yet far from inconceivable, see Slezak (2013).

9. For the same reason, we tend to distinguish only between language and metalanguage (or a language that speaks about a language) and also avoid using meta-metalanguage (a language that speaks about a language that speaks about a language: imagine explaining in English how one can translate a French sentence into Italian) as a redundant distinction, since any meta-chain can be reduced to a series of couples of object language and metalanguage, and this is sufficient to explain the interaction.

10. Logan (1944).

11. Fowler et al. (2008).

12. This is a rephrase of the famous view—advocated by the German philosopher Georg Wilhelm Friedrich Hegel (1770–1831) in his *Phenomenology of the Spirit* (Hegel 1979)—according to which what is rational is real and what is real is rational.

13. 'Graecia capta ferum victorem cepit et artes intulit agresti Latio.' (Conquered Greece took captive her savage conqueror and brought her arts into rustic Latium.) Horace, *Epistles*, book II, epistle 1, ll. 156–7, Horace (2011).

14. Nogee (2004).

15. Harrop and Das (2013).

16. Saponas et al. (2006).

17. Gardiner (15 August 2008).

18. Briere et al. (2013).

19. Gabbatt and Rushe (3 October 2013).

20. Clemens (2004).

21. *The Economist* (20 December 2005).

22. Doward and Hinsliff (24 February 2008).

23. BBC News (11 October 2007).

24. Norris (2001).

25. ALife is the scientific area of research that studies artificial life, e.g. in simulations and robots trying to recreate biological phenomena.

26. This is known as 'near field communication', a set of standards for ICTs that enables radio communication through touch or close proximity, to perform transactions or data exchange, for example.

27. 'From James Cameron to the Wachowski brothers to Steven Spielberg, US film-makers are paying homage to a groundbreaking Japanese anime—the movie that gave us today's vision of cyberspace', Rose (19 October 2009).

28. BBC News (9 August 2013).

29. Bridle (29 September 2013).

30. Dredge (27 June 2013).

31. BBC News (16 September 2013). Reed Hastings had expressed similar views before and is criticized by Mason (6 May 2013).

32. For a simple guide see Gardner and Krug (2006).

33. Uppenberg (2009).

34. For an accessible introduction see Barnatt (2010).

35. Barboza (9 December 2005), Thompson (25–31 March 2005).
36. Herzfeld (4 December 2012).
37. Dibbell (17 June 2007).
38. Dannenberg (2010).
39. Horowitz (2008), p. 80.
40. Luke (2013).
41. Anderson and Moore (2006).
42. Bond et al. (2013).
43. Pfanner (28 September, 2009), Ata et al. (2013).
44. Sauter et al. (1 February 2013).
45. The comparison is suggested by Klein (2000).

Chapter 3

1. Users still have Facebook ID numbers, which can easily be found by using online services such as <http://findmyfacebookid.com/>. Mine for example is 556011030.
2. The expression 'technologies of the self' was coined by the French philosopher Michel Foucault (1926–84), see Martin (1988).
3. Proust (1996), Overture.
4. Lenhart (19 March 2012).
5. Source: Informa, OTT messaging traffic will be twice volume of P2P SMS traffic this year, Press Release, 30 April 2013, available online.
6. C. Carter (12 September 2013).
7. James (1890), vol. i, pp. 239–43.
8. TED (Technology, Entertainment, Design) is a global set of conferences in which speakers are given only a few minutes (the maximum is 18, it is usually much less) to present their innovative ideas engagingly.
9. BBC News (7 June 2013).
10. Peter Steiner's cartoon was published in the New Yorker, 5 July 1993.
11. Plato presents his 'Allegory of the Cave' in Republic 514a–520a (see Plato 1997). Some people have spent all of their lives chained to the wall of a cave. They face a blank wall, on which they can see only shadows projected by things that are passing in front of a fire behind them. This is as close as Plato could get, technologically, to the idea of a virtual reality. Plato uses the analogy to explain how we mistake the world we perceive for the actual reality behind it. In the analogy, the philosopher is like a prisoner who escapes from the cave, realizes that the shadows on the wall are not real but merely stand for reality's true forms, and goes back inside the cave to help the other prisoners.
12. Giuseppe Verdi, La Traviata, Atto Primo, Preludio, Scena I, Coro I.
13. Plutarch, Theseus. The Internet Classics Archive, online.
14. Acts 7:58.

15. Thus, Ludwig Wittgenstein (1889–1951) was right in arguing that no private language (a language of my own, that only I speak) may subsist without a public language, but once a public language is available, the speaker may throw away the public language and privatize it, as it were. Wittgenstein was not denying that Hamlet could soliloquize in his own private language. He was denying that he could do so without appropriating a language that was public to begin with.

16. Wheeler (1990), p. 5.

17. Wolfe and Akenson (2005).

18. See Proust's famous episode of the involuntary memory retrieved by the taste of the madeleines in his *Remembrance of Things Past*.

19. Lacan is credited with having called attention to the importance of the phenomenon, which plays a central role in Foucault's philosophy, and in feminist theory.

20. Juniper Research (2012).

21. Heilman et al. (2011).

22. Fishman (2010).

23. Shin and Sneider (2011).

24. Berkman and Plutzer (2010).

Chapter 4

1. Pascal (2008), p. 347.

2. The letter is reprinted in Pascal (1997).

3. The machine could add. Subtraction was performed using complement techniques, in which the number to be subtracted is first converted into its complement, which is then added to the first number. Multiplications and divisions were then performed through series of additions or subtractions. Interestingly, computers employ similar complement techniques.

4. *Jetons* were tokens used as counters in calculation on a lined board similar to an abacus.

5. Pascal (1997), p. 59, my translation.

6. See now Hobbes (1991).

7. US Civil Service (1891).

8. Turing (1950).

9. The QWERTY keyboard, named after the first six letters on the top row of the keyboard, was devised in the early 1870s and made popular by Remington from 1878. It is suboptimal, because the layout was adopted in order to minimize mechanical clashes and jams when neighbouring keys were pressed simultaneously or in rapid succession. Despite this, its widespread use made it an ISO standard in 1972, see Noyes (1983).

10. The phenomenon of telepresence or presence at distance is not crucial in many contexts, such as surgery, where remote control and interactions are becoming widespread.

11. *Dead Souls* is a classic novel by the Russian writer Nikolai Gogol (1809–52). Published in 1842, is centred on Chichikov (the main character) and the people he encounters. The expression 'dead souls' has a twofold meaning. On the one hand, it refers to the fact that, until 1861, in the Russian Empire landowners were entitled to own serfs to farm their land. Serfs were like slaves: they could be bought, sold, or mortgaged, and were counted in terms of 'souls'. 'Dead souls' are serfs still accounted for in property registers despite their departure. On the other hand, 'dead souls' also refers to the characters in the novel, insofar as they have become fake individuals. For an English translation of Gogol's classic novel see Gogol (1998) in the Oxford World's Classics Series.

12. Van Duyn and Waters (7 August 2006).

13. Gogol (1998), ch. 5.

14. Raice (2 February 2012).

15. Source: *Financial Times* (13 June 2013), available online.

16. I was not the only one to be astonished, see Cohan (20 May 2013).

Chapter 5

1. See now Woolf (1999).

2. Turing (1936).

3. See now Woolf (2003).

4. Bentham (2011).

5. Asimov (1956).

6. The debate began in the late seventies, see for example Oldham and Brass (1979) and Sundstrom et al. (1982).

7. Warren and Brandeis (1890), p. 10.

8. Holvast (2009).

9. Lenhart and Madden (18 April 2007), p. iv.

10. Marwick et al. (2010), p. 13.

11. Orwell (2013).

12. Plato, *Republic*, II, 359b–360b. See Plato (1997).

13. A cookie is a tiny file sent from a website and stored in a user's web browser while the user is browsing it. Whenever the user visits the website again, the browser sends the cookie back to the server to notify the website of the user's previous activity. Cookies are designed so that websites remember information (e.g., items in a shopping cart) or record the user's browsing activity, including clicking particular buttons, logging in, or recording which pages were visited by the user in the past.

14. Soltani et al. (2009).

15. *The Economist* (16 December 1999).

16. Fineman and Mykitiuk (1994), see esp. the chapter by Elizabeth M. Schneider, 'The Violence of Privacy' (a reprint of her article published in 1990).

17. Warren and Brandeis (1890), p. 25, emphasis added.
18. Cohen (2000), p. 1426.
19. Warren and Brandeis (1890), p. 31.
20. Warren and Brandeis (1890), p. 33.
21. Quoted in Mills (2005).
22. Salinger (1951).

Chapter 6

1. Anderson (23 June 2008).
2. Plato, *Cratylus*, 390c; see Plato (1997).
3. Carr (1 July 2008).
4. Turing (1950), p. 442.
5. Turing (1950).
6. The full interview is available here: <http://www.youtube.com/watch?v=3Ox4EMFMy48>.
7. School of Systems Engineering, 'Reading to host Loebner Prize' available online at <http://www.reading.ac.uk/sse/about/News/sse-newsarticle-2008-05-16.asp>.
8. Chatterbots that imitate some everyday trite forms of conversations are old. The most famous of them, ELIZA, was created by Weizenbaum (1966), who became rather critical of AI, see Weizenbaum (1976).
9. Cosmides (1989).
10. Fiddick et al. (2000).
11. McCarthy (1997).
12. 'What is your aim in philosophy? To show the fly the way out of the fly-bottle.' Wittgenstein (2001), §309.
13. Dijkstra (1984).
14. Simon (1996).

Chapter 7

1. BBC News (2 December 2013).
2. Von Ahn et al. (2008).
3. On 2 December 2013, the highest reward was $28.69, and the lowest were $0.00 (volunteers) and $0.01.
4. Cited in Lohr (10 March 2013).
5. Fuggetta (2012). On Klout and the general phenomenon of influence see also Schaefer (2012).
6. Spartacus (*c.*109–71 BC) was a military slave leader in the Third Servile War, a major slave uprising against the Roman Republic.
7. Source: American Pet Products Association annual report, available online.

8. Source: Entertainment Software Association annual report, available online.
9. Wollslager (2009).
10. The standard reference is still Picard (1997).
11. Weizenbaum (1966).
12. Berners-Lee et al. (2001).
13. Source: W3C Semantic Web Frequently Asked Questions, available online.
14. In AI, the most difficult problems are called AI-complete or AI-hard whenever solving them requires making computers at least as intelligent as people, that is, when they presuppose the availability of strong AI. Examples of AI-complete problems usually include computer vision (see the example of CAPTCHAs in this book), understanding meaning, and dealing flexibly and succesfully with unexpected circumstances while solving any real-world problem. As of today, AI-complete problems can be solved by computer only with the help of human intervention.
15. O'Reilly (1 October 2005).
16. Source: ICT Data and Statistics Division, Telecommunication Development Bureau, International Telecommunication Union, *The World in 2013, ICT Facts and Figures*, available online.

Chapter 8

1. This is defined as the value of financial assets plus real assets (mainly housing) owned by individuals, less their debts.
2. Source: *The Credit Suisse Global Wealth Report 2011*, available online.
3. Source: *Nielsen Global AdView Pulse Q4 2011*, available online.
4. Source: Stockholm International Peace Research Institute, *Military Expenditure Database*, available online.
5. Source: PricewaterhouseCoopers, *Global Entertainment and Media Outlook 2007–2011*, available online.
6. Source: IDC, *Worldwide IT Spending Patterns: The Worldwide Black Book*, available online.
7. Using standard vocabulary, by nation I refer to a sociocultural entity comprising people united by language and culture. By state, I refer to a political entity that has a permanent population, a defined territory, a government, and the capacity to enter into relations with other states (Montevideo Convention, 1933). The Kurds are a typical example of a nation without a state.
8. The Latin word means 'estimate'. Already the Romans, who were well aware of the importance of information and communication in such a large empire for administrative and taxing purposes, carried out a census every five years.
9. Williamson (1993).

10. Anderson and Cavanagh (2000).
11. Williamson (2012). The expression 'Beijing Consensus' was introduced by Ramo (2004), but I am using it here in the sense discussed by Williamson (2012) and Halper (2010).
12. On volunteerism see Hockenos (2011); on digital activism, the Digital Activism Research Project (<http://digital-activism.org/>) offers a wealth of information.
13. I use the expression here in the post-Hegelian sense of non-political society.
14. The social space where people can meet, identify, and discuss societal problems, shaping political actions.
15. Rawls (1999).
16. For a more detailed analysis see Floridi (2012).
17. Source: *The New Atlantis* report, available online.
18. Source: Press release, *Digital Agenda: cyber-security experts test defences in first pan-European simulation*, available online.
19. *The Economist* (2 June 2012).
20. Floridi (1999).
21. For a study of how current international law applies to cyber conflicts and cyber warfare, see NATO Cooperative Cyber Defence Centre of Excellence (2013).
22. *The Economist* (7 June 2007).
23. Source: the *Wilson Quarterly*, report available online.

Chapter 9

1. Crutzen (2006).
2. Funabashi (2012).
3. Weaver (13 May 2010).
4. Glanz (22 September 2012).
5. *The Economist* (22 May 2008).
6. Glanz (22 September 2012).
7. Sterin (2013).
8. Boccaletti et al. (2008).
9. Webb (2008).
10. This consists in killing cells that divide rapidly, which include not only cancer cells but also cells in the bone marrow, digestive tract, and hair follicles.
11. Floridi (2013).

Chapter 10

1. I have sought to develop such an e-nvironmental ethics in Floridi (2013).

INDEX